Long Non-coding RNA

RNA Set

coordinated by
Marie-Christine Maurel

Long Non-coding RNA

The Dark Side of the Genome

Antonin Morillon

ISTE PRESS

ELSEVIER

First published 2018 in Great Britain and the United States by ISTE Press Ltd and Elsevier Ltd

ISTE Press Ltd
27-37 St George's Road
London SW19 4EU
UK

www.iste.co.uk

Elsevier Ltd
The Boulevard, Langford Lane
Kidlington, Oxford, OX5 1GB
UK

www.elsevier.com

Notices
Knowledge and best practice in this field are constantly changing. As new research and experience broaden our understanding, changes in research methods, professional practices, or medical treatment may become necessary.

Practitioners and researchers must always rely on their own experience and knowledge in evaluating and using any information, methods, compounds, or experiments described herein. In using such information or methods they should be mindful of their own safety and the safety of others, including parties for whom they have a professional responsibility.

To the fullest extent of the law, neither the Publisher nor the authors, contributors, or editors, assume any liability for any injury and/or damage to persons or property as a matter of products liability, negligence or otherwise, or from any use or operation of any methods, products, instructions, or ideas contained in the material herein.

For information on all our publications visit our website at http://store.elsevier.com/

British Library Cataloguing-in-Publication Data
A CIP record for this book is available from the British Library
Library of Congress Cataloging in Publication Data
A catalog record for this book is available from the Library of Congress
ISBN 978-1-78548-265-6

Contents

Foreword

The Modern RNA World

There is a strong case that the genetic programming of humans and other complex organisms has been misunderstood, because of the incorrect assumption that most genetic information is transacted by proteins. This assumption stems from studies of enzyme biochemistry and the lac operon in *E. coli*, in the middle of the 20th Century, consistent with the mechanical orientation of the age, and has persisted despite a number of surprises that should have given pause for thought.

The first was that genes in complex organisms are mosaics of protein-coding and non-coding sequences, the latter of which (termed "introns") were immediately and almost universally dismissed as evolutionary debris, a hangover from the early assembly of protein-coding genes, despite the fact that they are transcribed. The interesting alternative, that meaningful information is being transmitted by RNA, is equally if not more plausible.

The second surprise was that almost half of the human genome is comprised of retrotransposon-derived sequences, again dismissed as selfish genetic hobos, but the more interesting alternative is that these sequences are mobile regulatory cassettes.

The third surprise was that the number and repertoire of protein-coding genes is similar in nematodes and humans, despite orders of magnitude difference in their complexity. The ensuing rationalization was that the explosive potential of the combinatorics of "transcription factors" provides more than enough regulatory headroom to direct the ontogeny of a worm or a human.

This assumption, however, was not justified mathematically, by reference to decision theory, nor mechanistically, but has been accepted uncritically as it was comfortable. The discovery of small regulatory RNAs and RNA interference was treated as an add-on to the established protein-centric regulatory paradigm, especially since miRNAs regulate mRNA translation and stability, rather than being the tip of a regulatory RNA iceberg.

In contrast to protein-coding genes, the non-coding portion of the genome increases with developmental complexity, reaching over 98% in humans. Moreover the high throughput cDNA/RNAseq studies carried out over the past decade or so have shown that most of the mammalian genome is transcribed in a highly regulated fashion, producing, in addition to the small RNAs referred to above, a plethora of intronic and multi-exonic antisense and intergenic RNAs, collectively known as long noncoding RNAs (lncRNAs).

Some lncRNAs are precursors for small RNAs but most are highly cell type-specific, usually expressed in more restricted patterns than protein-coding genes, although there are exceptions. Some have questioned the relevance of lncRNAs because their sequences are not highly conserved, relative to those specifying proteins (although at least 20% of the mammalian genome is conserved at the level of RNA secondary structure), and many lncRNAs appear to be "lowly" expressed in RNAseq data.

The rapid evolution of lncRNAs (and indeed promoters) is not surprising given the different structure-function constraints of regulatory sequences and the likelihood that they are subject to positive selection for adaptive radiation. Moreover it is well accepted that, given the relatively stable core proteome, most adaptive radiation in animals is achieved by variation in the regulatory architecture that controls the patterns of gene expression rather than changes to the proteins themselves.

The perceived low expression of lncRNAs is a consequence of under-sampling of particular transcripts that are expressed in specific cells in complex tissues. *In situ* hybridization and high-resolution RNA sequencing have both shown that lncRNAs are highly precisely, that is uncommonly, expressed, not some sort of "transcriptional noise". Indeed, perhaps related to their cell specificity, lncRNA promoters are, on average, more highly conserved than those of protein-coding genes.

Although there is much to do to understand their full dimensions, it is clear that non-coding RNAs fulfill a wide range of functions in cell and developmental biology. There are many types of small RNAs, notably miRNAs referred to already, and pIRNAs that appear to control transposon mobilization. These have attracted a great deal of attention, but there are other less well-understood classes of small RNAs that derive from transcription initiation sites and splice junctions, and which may have a role in nucleosome positioning. Moreover, all H/ACA snoRNAs (from fission yeast to humans) produce miRNA-like molecules and all C/D box RNAs produce piRNA-sized fragments. tRNAs are also cleaved to produce specific fragments that are exported from cells, orthologs of which decorate the ends of some viral and human RNAs. There are regulatory worlds within worlds, and the functional links and networks among these small RNAs remains to be determined.

While most are not, some lncRNAs are widely expressed, like *Xist*, responsible for silencing one of the X-chromosomes in females. Another is *Malat1*, one of the most highly expressed RNAs in vertebrates, which is also chromatin-associated. Its function is unknown and its deletion produces only subtle developmental consequences. Another is *Neat1*, which is expressed and associated with enigmatic mammal-specific subnuclear organelles called paraspeckles in particular types of differentiated cells, and whose absence again produces only subtle phenotypes, mainly to do with placental reproduction. An interpretation I favor is that these RNAs are involved in setting the platforms for biology of learning.

An example is the highly expressed retrotransposon-derived brain RNA *BC1* whose deletion produces no overt developmental consequences but causes the loss of exploratory behavior – invisible in the cage but lethal in the wild. Another is the lncRNA *Gomafu*, which decorates modified spliceosomes in particular neurons, and has mechanistic links to schizophrenia. Others are associated with unknown double structures in the nuclei of Purkinje cells.

Many lncRNAs appear to be involved in determining cell identity. Most, but not all, are nuclear-localized and many are associated with chromatin-modifying complexes. This suggests that their prime function is to guide the 100 or so DNA and histone-modifying enzymes to differentially mark nucleosomes at millions of different places around the genome in different cells at different times during development. LncRNAs can also act as scaffolds for the assembly of DNA-RNA-protein complexes that organize chromatin architecture. Enhancers are transcribed in the cells in which they are active, thought to be a by-product of enhancer activation, but more likely to be involved in guiding the chromatin looping associated with enhancer action.

A surprisingly large number of lncRNAs are localized in the cytoplasm, with emerging evidence that some are involved in signal transduction processes. Others may create subcellular domains in the cytoplasm and in the nucleus, possibly interacting with intrinsically disordered regions in RNA binding proteins to create RNA granules or liquid crystalline regions. One lncRNA decorates a mysterious dumbbell domain in Purkinje cells.

The strength of RNA is its ability to span the digital and analog world in biology: it links 3-dimensional structures (formed by hydrogen bonds on the Watson-Crick face, the Hoogstein face and the ribose face through the 2'OH) that can interact with proteins, with sequences that can interact sequence-specifically with other RNAs and DNA. RNA was probably the primordial molecule of life, which transferred its analogue functions to the more chemically versatile proteins and its informational functions to the more stable and easily replicable DNA. It is likely that RNA underwent a rebirth, as the intermediary for the epigenetic processes that guide the development of complex organisms.

These considerations imply a modular structure, which is supported by recent evidence of universal splicing of lncRNA exons and the location of evolutionary conserved structures within exons. The observations that alternatively spliced exons are localized with promoters and that exons are preferentially located in nucleosomes suggest that lncRNA exons may not only be the modular unit of structure-function, but also of histone-based epigenetic regulation. Parsing the structure-function relationships in lncRNAs is a big challenge that will be made much easier if regulatory RNA sequences are truly modular – in which case, once recognized, these modules may form the basis of a new Rfam, like the Pfam database that has been so useful in identifying orthologous domains in proteins.

There are many mysteries in RNA biology, two in particular. The first is the expression of 3'UTRs separate from their normally associated protein-coding sequences, with genetic evidence that they can transmit information in trans. 3'UTRs are well established to control mRNA translation and half-life through cis-acting protein and small RNA interactions. Why they should have evolved trans-acting functions is unknown, but a clue may be gleaned from the fact that 3'UTRs have expanded greatly during vertebrate evolution, and in many cases are longer and/or more highly conserved than their associated protein-coding sequences.

The second is the almost 14,000 ultraconserved elements of 100 bp or more, which evolved rapidly in tetrapod evolution, and then froze in the amniotes, being almost identical in all mammals. These non-coding sequences are transcribed, but cannot be explained, either in evolutionary or molecular terms. Some have been deleted in mice, but show no overt developmental phenotype, which suggests they may have another (very important) role in birds and mammals, likely parenting and learning. In any case, until these non-coding elements and their mechanism of action are understood we will not understand human biology or evolution.

The new frontier is to understand the role of RNA editing and the 140 or so different modifications of RNA, coined the epitranscriptome. Adenosine to inosine editing expands massively with cognition, especially in primates, on the back of 3 waves of colonization of the primate lineage by Alu elements. The APOBEC enzymes, which catalyze C>U editing, are vertebrate-specific, expand in mammals especially primates where one family (which appears to be involved in regulating retrotransposon mobilization) exhibits strong signatures of positive selection in human evolution. The lack of enzymes that catalyze RNA modifications leads to various diseases, notably intellectual disability and cancer.

There is also emerging evidence of RNA-mediated transgenerational epigenetic inheritance, so it may be that RNA has also been coopted to enable plasticity across generations. The presence of reverse transcriptases, especially in the brain, also implies a much more vibrant interplay between DNA and RNA, both in real and evolutionary time.

In conclusion, what was dismissed as junk because it was not understood likely holds the key to understanding human development and human intelligence. RNA is not simply a passive and ephemeral intermediate between gene and protein, but the computational engine of cell biology, development, the brain and probably evolution itself.

John S. MATTICK
Garvan Institute of Medical Research
and University of New South Wales
Sydney, Australia

Preface

98% of the human genome does not code for proteins but represents what is commonly called black matter, or the dark side, of the genome. It is now admitted that this dark side is expressed as non-coding RNA whose numbers are increasing with the evolution of sequencing technology and its sensitivity. Although the number of genes coding for proteins has remained stable over the last few years at around 20,000 genes in humans, the number of non-coding genes has reached several tens of thousands. Several examples, characterized by different laboratories around the world, have turned out to be fundamental to the regulation of gene expression and have major impacts on cell development and tumor progression in cancer. A large international community of scientists has been established and the interest raised by these long non-coding RNA has only increased alongside the idea of understanding the epigenetic regulation of the genome.

In this book, I put forward a non-exhaustive state of our current knowledge on long non-coding RNA. In Chapter 1, I place their discovery in the context of the history of molecular biology and discuss, in Chapter 2, today's suggested nomenclature for grouping them into families. Despite all the efforts made over the past decade, it still remains impossible to predict the function of a non-coding RNA based solely on its primary sequence. However, in different model organisms, a certain number of them have been characterized molecularly, biochemically and genetically and have revealed crucial functions essential to gene expression, maintenance and genome integrity. In Chapter 3, I aim to describe a few examples to encourage jumping into their respective literature, which only gets denser by the month.

Finally, in the last two chapters I describe in more detail the roles of specific long non-coding RNA in cell and organism development, but also in the control of tumor progression as these two aspects often bear commonalities. I aim to give the reader the keys to understanding these long non-coding RNA in these two lines of research.

The subject of cancer also provides an opportunity to underline that the high singularity of these long non-coding RNA, in terms of each tissue or cell type, making them excellent biomarker candidates for diagnostics or prognostics. The applications are wide and varied with certain studies now dedicated to exploiting them as therapeutic targets.

With this book, I hope that each reader finds a framework to enable them to explore, where needed, the variety of literature that I have not been able to cite in full, and to motivate students and researchers to bring their attention and eventually illuminate this dark side of genomes through their present or future research.

There is also a glossary and list of abbreviations at the end of the book to make it more accessible to students and researchers who want to further explore the immense quantity of data generated by this new continent.

Antonin MORILLON
October 2017

1

Non-coding RNA, Its History and Discovery Timeline

The complexity of eukaryotic transcriptomes and the rapid development of high throughput sequencing technology have led to an explosion in the number of long non-coding RNA (or lncRNA) recently identified and as yet undescribed. Current challenges include not only their precise labeling but also their functional characterization and the clinical pertinence of these biological objects. To begin with, it is important to return to the characteristics of RNA, a molecule essential to cellular functionality.

The timeline of discoveries linked to non-coding RNA is shown in Figure 1.1 and its history will be detailed in the first half of this chapter, from DNA to the first non-coding transcripts. Next, the role of global genomic and transcriptomic studies in changing our vision of RNA's capacity in gene expression regulation circuits and cellular complexity will be discussed. This functional diversity has given rise to various extensive classifications of lncRNA.

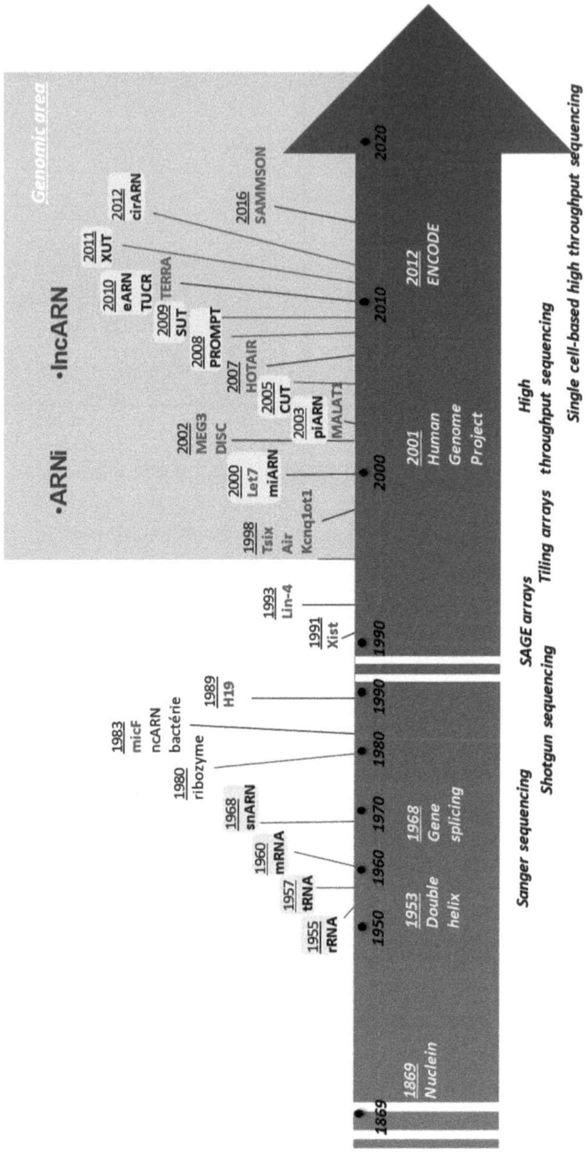

Figure 1.1. *Timeline of the main discoveries of RNA biology and in particular of eukaryotic non-coding RNA. For a color version of this figure, see www.iste.co.uk/morillon/RNA.zip*

NOTES ON FIGURE 1.1.– In light pink are the discoveries of large non-coding RNA families. In red are typical examples of specific non-coding RNA. In yellow are a selection of the big discoveries in molecular biology. Beneath the timeline are the novel technologies that enabled the characterization of non-coding RNA families. In violet are the discoveries of mechanisms regulated by non-coding RNA.

1.1. The biology of RNA, a century of history

The defining and conceptualization of RNA molecules in cell biology date back to 1869 and the discovery of nucleic acids. It took more than a century for researchers to finally identify non-coding transcription and to begin to suggest regulatory possibilities.

1.1.1. *From nuclein to the double helix*

At the end of the 19th Century, several essential discoveries foreshadowed the era of molecular biology. Although Friedrich Miescher isolated the contents of the cell nucleus, which he named "nuclein" (containing nucleic acid) (Dahm, 2005) in 1869, the interests of scientists at the time focused on the proteins thought to carry genetic information. It was only in 1944 that the link between nucleic acid (of the nucleus) and genetic information was made when Oswald Avery suggested that DNA was the carrier of genetic information (Avery *et al.*, 1944). As such, the association between the Mendellian genetic model and Miescher's "nuclein" was missing for more than half a century.

Following the visualization of the double helical structure of DNA by James Watson and Francis Crick in 1953, it was suggested, in 1961, that RNA might provide the intermediary molecule in the flow of information between

DNA and proteins (Cobb, 2015). Outlined in 1958 by Francis Crick and then by François Jacob and Jacques Monod, the central dogma of molecular biology included gene transcription of DNA to RNA in the nucleus followed by protein synthesis. It was also confirmed that the flow of information was only from DNA to RNA, then from RNA to the protein and never the reverse (Cobb, 2015). The mediator role of RNA has become a central focus for research, essential to the development of modern molecular biology.

1.1.2. *The "RNA world" concept*

It was only in 1955 that Georges Palade identified the very first non-coding RNA (ncRNA) that made up part of the most abundant RiboNucleoProtein (RNP) complex of a cell: the ribosome. Soon after, in 1958, a second class of ncRNAs were discovered by Mahlon Hoagland and Paul Zamecnik for their role as intermediaries between amino acids and RNA: transfer RNA (tRNA).

In 1960, François Jacob and Jacques Monod defined "messenger RNA" (mRNA) as the intermediate molecule carrying genetic information for protein synthesis. Following this, Crick and his team established that the genetic code is a universal non-overlapping triplet code in which three nucleotides code for one amino acid (Crick, 1968).

The discovery of heterogenous nuclear RNA (hnRNA) in the late 1960s led to the study of rRNA maturation and the discovery of splicing (Berk, 2016; Lewis *et al.*, 1975; Weinberg and Penman, 1968; Zieve and Penman, 1976).

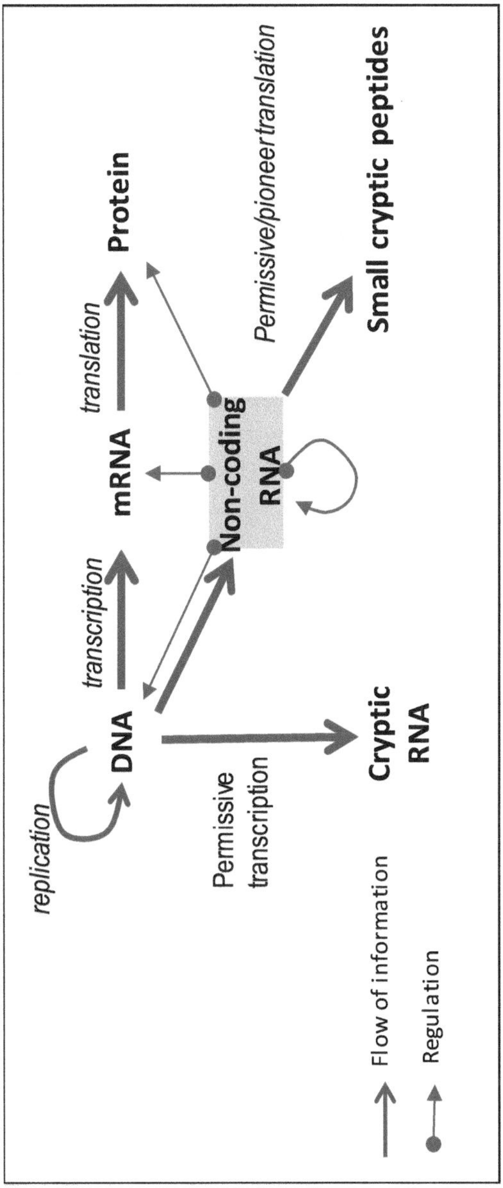

Figure 1.2. *Flow of genetic information and its regulators. For a color version of this figure, see www.iste.co.uk/morillon/RNA.zip*

The regulators in red in Figure 1.2 were not known in 1958 during the formulation of the initial dogma of information flow, nor were the permissive transcription and translation processes.

Although Jacob, Monod and Crick had all stated that RNA was not just a messenger, for a long time many scientists treated it as a simple unstable intermediate molecule, neglecting the potential active roles of other classes of ncRNA. However, this concept was partially changed in 1980 when Thomas Cech and Sidney Altman discovered that RNA molecules could also act as catalysts for chemical reactions such as self-splicing (Kruger *et al.*, 1982) or RNA degradation through ribonucleotide nuclease P, or RnaseP (Guerrier-Takada *et al.*, 1983). These RNA enzymes, or ribozymes, have since been acknowledged as key actors in the flow of genetic information (Figure 1.2), as part of both the ribosome and the spliceosome (Butcher, 2009; Cech, 2000).

The discovery of ribozymes led to the hypothesis of the "RNA world" which proposed that pre-biotic life revolved around RNA. Further studies of its roles in cell biology have revealed that RNA is necessary for DNA replication and that ribonucleotides are precursors of DNA's deoxyribonucleotides.

Furthermore, as previously mentioned, RNA plays an important role in all stages of protein synthesis whether as a template (mRNA) or as an actor (ncRNA: rRNA, tRNA, etc.) (Bernhardt, 2012). The latter are constitutively expressed in the cell and are necessary for vital cell functions. These maintenance ncRNAs are the subject of many specialized publications and will not be explored in this book. Other classes of regulatory ncRNA were discovered in the 1990s and will be broadly discussed here. The origin of those non-coding RNA and the fact they constitute an additional layer of genetic expression started to be discussed at this date.

Several hypotheses were raised, especially the interconnections with introns (Mattick, 1994; Morris and Mattick 2014). The origin and extension of the "RNA world" concept was increasingly acknowledged. These ncRNAs are expressed very specifically throughout the stages of embryonic development, in certain tissues or pathological states and also play multiple roles in the regulation of gene expression and genomic stability.

1.1.3. *Small bacterial RNA: pioneers of non-coding RNA*

The very first ncRNA regulator defined was micF, from the Escherichia coli bacteria. It was described in 1987 by Masayuki Inouye and his team (Inouye and Delihas, 1988) as the first RNA that regulates the expression of a gene through sequence complementarity, and represents the main class of bacterial small ncRNA regulators (sRNA). It was shown that the ncRNA of micF suppresses the translation of a target mRNA coding for a porine (outer membrane protein F, OmpF) that is involved in passive transport across cell membrane. When micF transcription is activated, it inhibits the expression of the ompF gene both at the mRNA level and within the protein.

Following the *in vitro* characterization of RNA's duplex structure, it was shown that micF binds to the ribosome binding site (RBS) of the ompF mRNA on the Shine-Dalgarno sequence, thus inhibiting the translation of ribosomes (Figure 1.3).

More recently, it has been shown that the regulation of gene expression by micF extends to other genes, including the mRNA lrp (Leucine responsive protein) (Corcoran *et al.*, 2012). Lrp is a transcription factor that regulates the expression of genes in the E. coli bacterium in reponse to osmotic changes and the availability of nutrients. Remarkably, Lrp regulates the expression of micF and vice versa, thus creating a feedback loop and proving the important role of micF in the overall regulation of genes and metabolism. Similar mechanisms have been found in the Salmonella bacteria, indicating a conservation of this regulatory pathway through evolution (Delihas, 2015). Since then, many other sRNA, whose lengths vary from 50 to 500 nucleotides (nt), have been discovered, including ncRNAs that act in *trans* or *cis*, RNA thermometers and riboswitches. They all work in pairs, thus inhibiting the translation of targeted mRNAs and inducing their degradation.

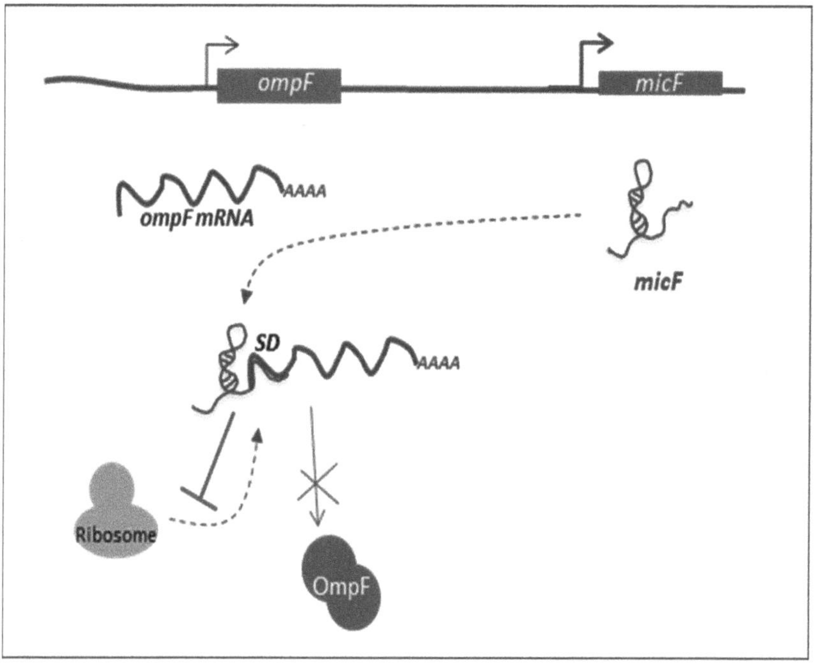

Figure 1.3. *Regulation of ompf expression by the non-coding RNA micF. For a color version of this figure, see www.iste.co.uk/morillon/RNA.zip*

NOTES ON FIGURE 1.3.– micF partially binds to the 5' region of the ompF mRNA blocking recognition of the Shine-Dalgarno (SD) site by the ribosome, therefore also blocking the translation of mRNA into the OmpF protein.

1.1.4. *Micro-RNA and RNA interference*

In the early 1990s, through co-expression of transgenes or viral infection experiments in different eukaryotic organisms, several scientists observed a phenomenon of protein synthesis inhibition mediated by RNA. The regulatory effects of these molecules would dramatically change the perspective of RNA as a simple messenger. Initial studies described the phenomenon as "co-suppression" in plants, as "post-transcriptional gene silencing" (PTGS) in nematodes or as "attenuation" in fungi, but none of them suspected it to be RNA itself. The first micro (mi)RNA was characterized in the Caenorhabditis (C.) elegans nematode in 1993 by Victor Ambros and his colleagues. Ambros discovered that the lin-4 gene produced small RNAs of 22 and 61 nts from a longer non-coding precursor. The precursor RNA forms a stem-and-loop structure that is cut to create shorter RNA strands with an antisense complementarity in the untranslated 3' region (UTR: Un-Translated Region) of the lin-14 transcript (Lee *et al.*, 1993). The pairing of lin-4 RNA with lin-14 mRNA was proposed as a molecular "post-transcriptional gene silencing" mechanism, thus decreasing the levels of LIN-14 protein in the initial larval stages of nematode development (Wightman *et al.*, 1993). Michael Wassenegger observed a similar phenomenon in plants that he described as "homology-dependent gene silencing". This process is mediated by the incorporation of viral RNA that induces the methylation of viral cDNA and the silencing of the gene expression (Wassenegger *et al.*, 1994). The whole process of "gene silencing through RNA" was put forward in 1988 by Andrew Fire and Craig Mello in similar experiments with the *C. elegans* unc-22 gene (Fire *et al.*, 1998).

In 2000, let-7, another essential miRNA, was identified in *C. elegans* and as homologs in several other organisms including humans (Ameres and Zamore, 2013; He and Hannon, 2004). Biogenesis, as well as the molecular mechanisms of regulation through miRNA were largely characterized by Thomas Tuschl. His team showed that in *C. elegans,* a long double-stranded RNA is transformed into shorter 21-25 nts fragments. Next, it was shown that the premature transcripts in the nucleus are transformed into hairpin-shaped RNA by the microprocessor complex containing the endonuclease protein, Drosha, before being exported to the cytoplasm where they are cleaved into small double strands by the type III RNAse protein, Dicer. One of the strands passes through the RISC complex and is then directed through complementarity to an mRNA molecule, thus inducing a repression in translation (He and Hannon, 2004). This simplified schematic (Figure 1.4) constitutes the basis of the interference mechanism of RNA (RNAi) and actually combines all the gene silencing phenomena mediated by small ncRNA including miRNA, small interfering RNA (siRNA) and Piwi-interacting RNA (piRNA) at the transcriptional (TGS) and post-transcriptional levels (PTGS) (Castel and Martienssen, 2013; Montgomery, 2004).

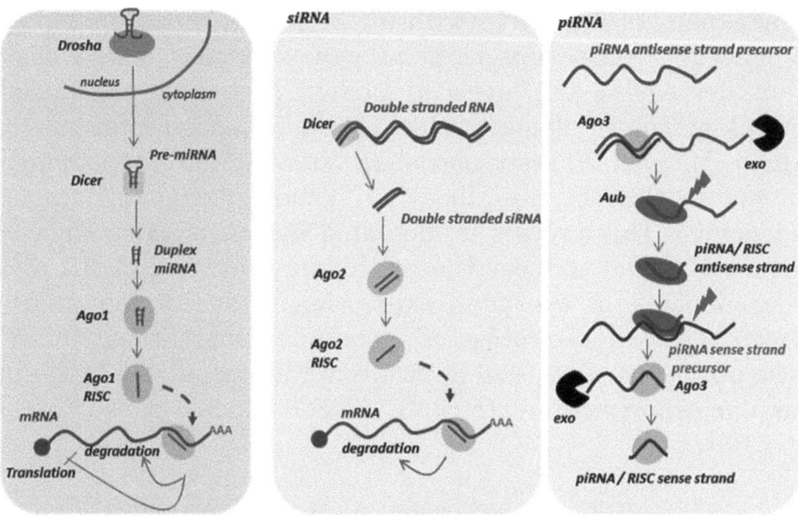

Figure 1.4. *Model of RNA interference. For a color version of this figure, see www.iste.co.uk/morillon/RNA.zip*

NOTES ON FIGURE 1.4.– Three modes of maturation of RNA miRNA, siRNA and piRNA. For the miRNA, in the nucleus, Drosha cleaves the stem-loop of pre-miRNA RNA which is then exported into the cytoplasm where DICER matures the loop and generates a duplex miRNA taken up by AGO1 that unravels it into a single strand RNA. The AGO1/RISC complex pairs with a target mRNA to prevent its translation and/or initiates its degradation. For siRNA, the double-stranded RNA is cleaved by DICER, generating small double-stranded RNA which is taken up by AGO1 to unravel it into a single strand. The AGO2/RISC complex targets an mRNA and initiates its degradation. For piRNA, a single strand precursor is recognized by an antisense strand carrying AGO3 which protects the double strand from degradation with an exonuclease. AUB cleaves the remaining single strand and forms the piRNA/RISC precursor. This targets an antisense RNA for cleavage while AGO3 protects the single strand from complete degradation through an exonuclease and forms the piRNA/RISC sense strand. This return phenomenon is called "ping-pong".

Although RNA interference resulted in a breakthrough for modern biology and biotechnology and the emergence of the concept of non-coding RNA as a regulator, the pertinence of the role of lncRNA remains largely unknown. However, some lncRNAs such as H19 and Xist were studied in the late 1980s and led to milestones in dosage compensation mechanisms in mammals. Dosage compensation, also called genomic imprinting or parent-of-origin specific expression, is fundamental to the diversity of gene expression in a lineage. Essentially, the parental alleles of the chromosomes inherited from the male or female are not always expressed in the same way. Certain domains inherited from the mother or father are switched off or overactive depending on a marker left on the DNA in the region inherited from the parent. The fundamental discovery of Xist and H19 put lncRNA where it is now, as a key factor in the establishment and maintenance of the parental imprint.

1.2. The discovery of long non-coding RNA in the pre-genomic era

In the 1980s, scientists used differential hybridization methods of cDNA libraries to clone and study non-coding genes. Later, an *a posteriori* approach was adopted without regard for the coding potential of RNA. Thanks to this approach, H19 became the first non-coding gene discovered even though at the time, it was initially classed as mRNA (Pachnis *et al.*, 1984).

1.2.1. *H19: the first in the history of long non-coding RNA*

The detailing of the genomic imprinting phenomenon, of parent-of-origin specific gene expression, forms a part of the dosage compensation mechanisms that work by sequestering genes in an epigenetic manner (see glossary). Independently, two "imprinted" genes were identified: Igf2r coding for the paternally expressed protein and H19, expressed by the mother (Figure 1.5). The two genes were located on chromosome 7 in mice, near the H19/IGF2 cluster (Barlow *et al.*, 1991; Bartolomei *et al.*, 1991). What makes H19 unusual is the absence of its translation into a protein, even though the gene contains small open reading frames. H19 is conserved in mammals and the transcript, in abundance, presents mRNA characteristics: transcribed by polymerase II RNA, spliced, polyadenylated at the 3' end and exported to the cytoplasm (Brannan *et al.*, 1990). The heterologous expression of H19 in transgenic mice proved to be fatal at prenatal stages, which not only suggested that the dosage of this lncRNA was strictly controlled, but that it also played an important role in embryonic development. However, the function of H19 as an RNA molecule remained a mystery until the functional characterization of another lncRNA involved in the dosage compensation of the mammalian X chromosome, Xist, was defined. Since then, H19 has been the subject of extensive study and represents the prototype of an lncRNA with multiple functions.

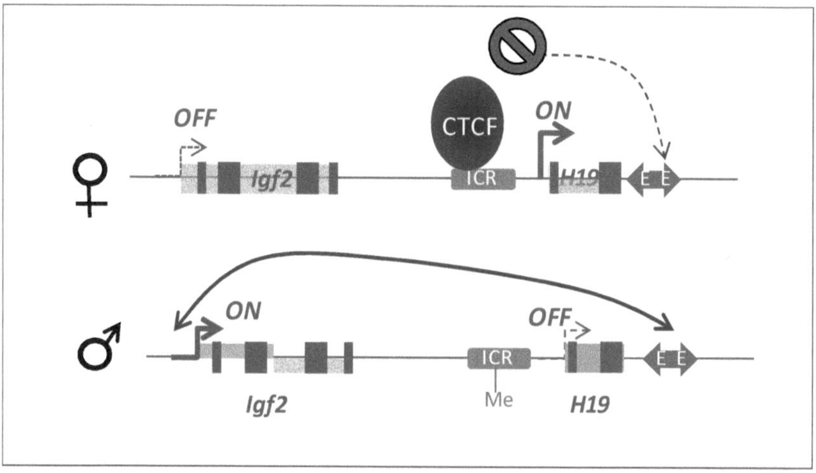

Figure 1.5. *Regulation of parental imprinting and expression of H19. For a color version of this figure, see www.iste.co.uk/morillon/RNA.zip*

NOTES ON FIGURE 1.5.– On the maternal allele, binding of the CTCF repressor on the ICR region prevents the Igf2 activator sequence enhancers (E) from touching the IGF2 promoter. The maternal gene is thus switched off and the lncRNA H19 is activated. The opposite is seen on the paternal allele where the methylation of the ICR DNA (Me), inherited from paternal chromosomes, prevents the CTCF factor from binding and thus allows the activation of the IGF2 gene by enhancer sequences. The methylation of ICR represses H19. Paternal IGF2 is therefore switched ON.

1.2.2. *Inactivation of X, the eXISTence of XIST*

In living organisms, gender can be determined in many ways and is defined in mammals by the X and Y chromosomes: whereas males have a single copy of the X and the Y chromosomes, females have two X chromosomes in their karyotype. The X chromosome, although involved in male and female gender differentiation, carries many genes whose functions are not a part of gender determination. As a consequence, there is a need for dosage compensation between males and females to enable these genes to be expressed in an equivalent manner.

In Drosophila, the dosage compensation mechanism consists of doubling the transcription of the single X chromosome in males. In mammals, one of the two X chromosomes in females is inactivated. This phenomenon, called X Chromosome Inactivation (XCI), was first discovered in mice by Mary Lyon in 1961 (Lyon, 1961) and later generalized to other mammals. XCI is established in the first stages of development and initiated by a single locus, the X inactivation center (Xic).

In the early 1990s, it was shown that this locus produced a long non-coding RNA, XIST (X-inactive-specific-transcript), weakly expressed in both the male and female undifferentiated embryonic stem cells (ES). During differentiation, the expression of XIST is randomly activated by a single allele, in female cells, by the future inactivated X (Xi). Retained in the nucleus, XIST sets off gene silencing in *cis* (Figure 1.6), physically placing itself and spreads widely on the future Xi (Borsani *et al.*, 1991; Brockdorff *et al.*, 1991; Brown *et al.*, 1991).

Unlike H19 and other lncRNA involved in dosage compensation, XIST is very unusual because it leads to silencing in the whole chromosome. The spread of XIST along Xi, called “coating”, requires the enveloping of RNA around the X and the recruitment of multiple factors, including the repressor complexes Polycomb 1 and 2 (PRC1 and PRC2). These trigger a cascade of changes in the chromatin and a spatial reorganization of the entire Xi before finally resulting in the stable repression of nearly all the genes linked to Xi throughout the course of development and adult life (Gendrel and Heard, 2014). While the expression of XIST is crucial for the initiation of XCI, in somatic cells XIST and Xic are not indispensable for the maintenance of gene silencing in mice (Brown and Willard, 1994). XIST does not only play a role in stable inactivation (Heard *et al.*, 1999). Xic was first defined in mice as the minimum region on the X chromosome that contained all the necessary and sufficient sequences for the initiation of XCI. Xic is spread over a million base pairs and transcriptomic studies have revealed that this region contains several coding and non-coding genes, including Linx, Ftx and others. Some non-coding genes in Xic show poor sequence conservation between man and mouse, including the XIST sequence itself (Chureau *et al.*, 2002). Tsix in particular, characterized in mice, is an antisense transcript that overlaps the whole XIST gene (Figure 1.6).

In humans, the main Tsix regulator elements are truncated and the transcript only overlaps XIST at the 3' end. These differences remove Tsix's role in the transcriptional repression of XIST on the future X chromosome that is active in humans (Lee and Lu, 1999; Migeon *et al.*, 2002). Recently, another lncRNA, Xact, has been discovered in human ES cells. This gene is located in the intergenic region, outside of Xic, and is not conserved in mice. In human female stem cells, Xact is expressed and covers both X chromosomes. This lncRNA appears specific to pluripotent cells and may ensure a specific control of XCI in humans (Vallot *et al.*, 2013). Over the last few years, the biogenesis of XIST, its structure and the molecular mechanism of XCI have been at the heart of many development studies in different mammals (Gendrel and Heard, 2014).

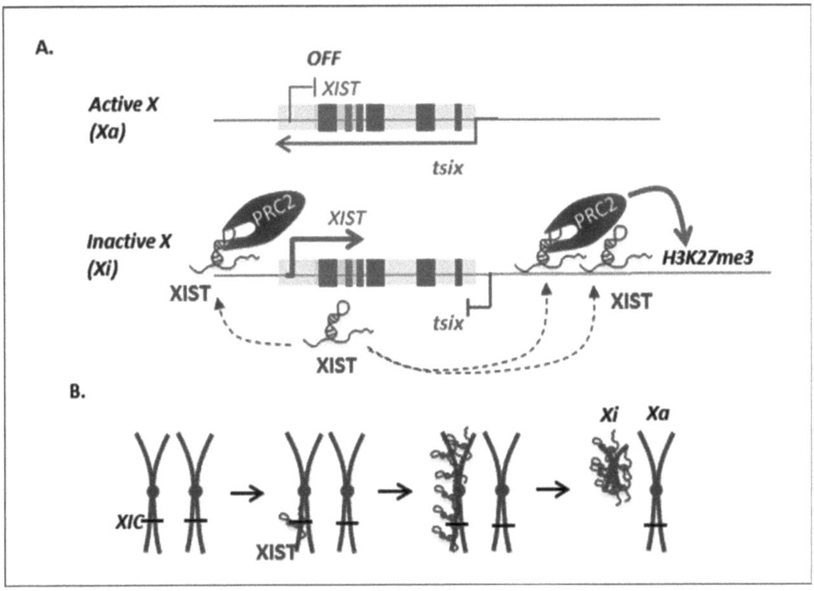

Figure 1.6. *Regulation of the inactivation of the X chromosome by XIST. For a color version of this figure, see www.iste.co.uk/morillon/RNA.zip*

NOTES ON FIGURE 1.6.–

A. The active X chromosome (Xa) transcribes the XIST antisense (Tsix) which represses the expression of XIST. The inactive X (Xi) shows a weak rate of transcription of Tsix and XIST is activated. It spreads along the length of the X chromosome, recruits/activates PRC2 which deposits the H3 lysine 27 trimethyl histone (H3K27me3), a marker of chromatin repression. The large majority of the inactive X chromosome genes are thus repressed..

B. Stages of X chromosome inactivation. XIST is expressed on the X inactivation center (XIC) on one of the 2 chromosomes, spreads and inactivated the whole chromosome (Xi).

These pioneering studies on H19 and XIST revolutionized our vision of non-coding RNA functions and the biological pertinence of lncRNA in general. It has been suggested in particular that non-coding RNA could provide a link between the genetic code and the epigenetic code to modulate genome accessibility in order to finely regulate its expression throughout both its cellular life and development (Figure 1.7). These examples have shown the complexity and versatility of regulation circuits orchestrated by a single lncRNA. They have also stimulated further discoveries and suggestions of potential mechanisms for other non-coding transcripts that remain uncharacterized. A global effort in the identification and characterization of lncRNA was launched in the 2000s with the complete sequencing of the human genome and through the ENCODE (Encyclopedia of DNA Elements) project (Djebali *et al.*, 2012).

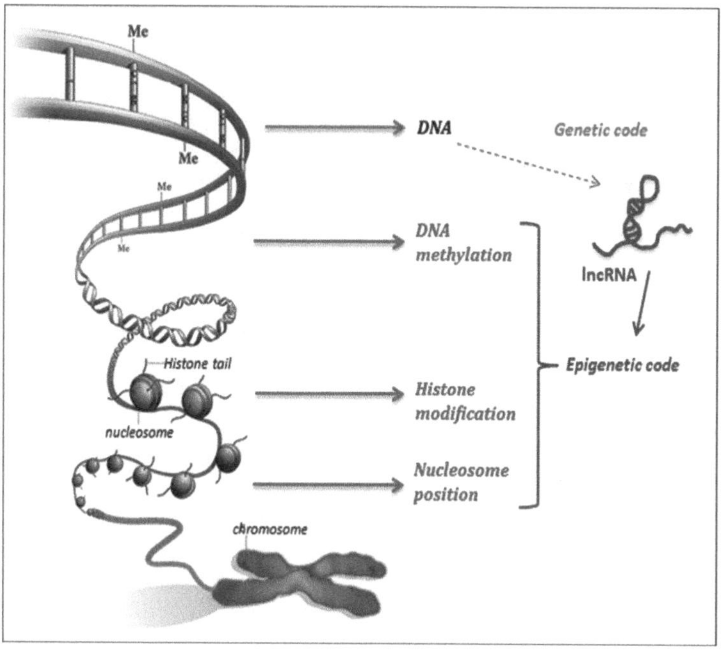

Figure 1.7. *Non-coding RNA, the link between the genetic and epigenetic code. For a color version of this figure, see www.iste.co.uk/morillon/RNA.zip*

NOTES ON FIGURE 1.7.– Through sequence complementarity, non-coding RNA can place epigenetic markers on the DNA sequence by modulating, targeting, and activating DNA modification complexes (Methylation, histone position or post-transcriptional modification).

1.3. From the non-coding genome to the non-coding transcriptome, the advent of the genomic era

Our modern vision of eukaryotic transcriptomes was preceded by in-depth studies of genomic DNA and the discovery that, besides protein coding (PC) sequences and the essential regulator elements for the transcription of PC genes (PCG), a large majority of the genome contained sequences considered as evolutionary fossils that are, therefore, useless. For about 20 years, to differentiate these sequences from PC sequences, these regions of DNA were called non-coding and considered as “selfish” DNA (often truncated repeat elements) or “undesirable” (Orgel and Crick, 1980).

1.3.1. *The human genome project: genomic DNA is essentially non-coding*

In 1978, Frederick Sanger generated the very first complete genomic sequence : that of the viral genome of the ϕX174 bacteriophage (Sanger *et al.*, 1978). Since then, Sanger sequencing has been widely used across the world. Its discovery and development earned Sanger and Walter Gilbert the Nobel Prize for chemistry. Throughout the following years, several viral genomes were sequenced and in the late 1990s, a complete human genome sequencing project, the Human Genome Project (HGP), was launched by the National Health Institute (NIH, United States). In parallel, the American biochemist and entrepreneur Craig Venter founded his own company to reach the same goal and thus the race to unravel the human genome began. The first bacterial genome was published in 1995 (Fleischmann *et al.*, 1995), followed in 1999 by the sequencing of the euchromatic part of the human chromosome (Dunham *et al.*, 1999) which covers around 65% of what is now known as chromosome 22. Finally, a first copy of the complete human genome was published in *Nature* in 2001, covering 96% of the euchromatin (Lander *et al.*, 2001b), followed the very next day by Craig Venter's publication in *Science* of the whole genome sequence (Venter *et al.*, 2001). Since 2003, regular updates have completed these initial sequences. At the same time, the genomes of several other organisms were obtained, notably that of yeast, the *pufferfish* fish, the *C. elegans* worm, the *D. melanogaster* fly and the mouse (Adams *et al.*, 2000; Goffeau *et al.*, 1996; Roest Crollius *et al.*, 2000; Waterston and Sulston, 1995; Waterston *et al.*, 2002), allowing for advanced comparative studies to be carried out.

The first surprise to come out of this complete genomic sequencing effort was the number of rather weak PCG compared with what was expected. Initial studies predicted 70,000 to 80,000 coding genes in the human genome (Antequera and Bird, 1993), and a figure closer to 100,000 genes was suggested in the mid-1980s. The whole sequencing results were predicted to be around 31,000 PCG in 2001, which was then decreased to 22,287 PCG in 2004 (consortium, 2004; Lander *et al.*, 2001b) and 19,600 is suggested in the current Gencode V25 version (Wright *et al.*, 2016). Overall, the results are astounding: gene coding exons represent only 1.2% of the human genome, while 24% and 75% are thought to be introns or non-coding intergenic DNA (Figure 1.8). This, however, is not the only surprise. By comparing the proportions of coding and non-coding genome, it became apparent that the percentage of non-coding evolves spectacularly with the "complexity" of the organism, even if the notion of complexity itself remains to be defined (Figure 1.9). Either way, it is striking to consider that 25 to 50% of bacterial or unicellular eukaryotic genomes are non-coding whereas in vertebrates this figure lies at over 95%. The idea that the non-coding genome is somehow involved in specialization and cell identity therefore begins to pave its way.

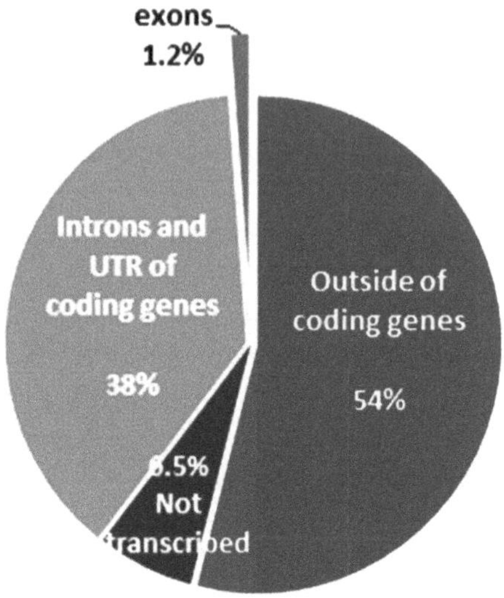

Figure 1.8. *Non-coding genome. Proportion of transcribed and non-transcribed regions in genes coding for proteins (exons) and non-coding sequences (introns, UTRs and outside of genes) in the human genome, according to ENCODE. For a color version of this figure, see www.iste.co.uk/morillon/RNA.zip*

1.3.2. *Permissive transcription and the hidden face of the genome*

Although the idea of genome fossilization (through inert strata of non-coding genes) continued to be discussed, the notion was shattered when the HGP revealed that the majority of the genome is transcribed, whether protein-coding or not. The use of oligonucleotide sensors covering human chromosomes 21 and 22 had previously revealed that 90% of the cytosolic polyadenylated transcripts detected corresponded to non-coding genomic regions and not to exons (Kapranov *et al.*, 2002). Similar results were revealed by the FANTOM and RIKEN consortia following the analysis of human and mouse transcriptomes (Carninci *et al.*, 2005; Okazaki *et al.*, 2002). More than 60,000 complete mouse cDNA were sequenced to generate precise maps of the 5' and 3' ends of all transcripts, thus defining the transcription start sites (TSS) and termination sites (TTS). Remarkably, more than 23,000 ncRNA resulting from the transcription of both sense and antisense genes represented around two thirds of the mouse genome (Katayama *et al.*, 2005). For the first time, antisense transcription was suggested as a contributor in gene expression regulation at the transcription level in mammals (Katayama *et al.*, 2005).

These results were later confirmed by further in-depth studies on human genomes carried out by the ENCODE (Encyclopedia of DNA Elements) Consortium. In its pilot phase, this project involved more than 200 experiments (Birney *et al.*, 2007) and subsequently up to 1,640 databases from 147 different cell lines were defined (consortium, 2012). Thanks to various sequencing techniques, the landscape of DNase I hypersensitive sites, histone modification sites, transcription factor binding sites and collection of transcriptomes has been defined. These studies concluded that around 93% of the human genome is actively transcribed and associated with at least one primary transcript (coding and non-coding exons and introns). Of these transcripts, around 39% of the genome represent PCGs (of the poly-A promoter), 1% are coding exons while the remaining 54% are not PCG (Figure 1.8). More recently, the mouse counterpart of the ENCODE Consortium confirmed previous studies by publishing a similar analysis showing that 4 to 6% of the mouse genome produces mRNA whereas at least 87% of its genome is transcribed (Stamatoyannopoulos *et al.*, 2012; Yue *et al.*, 2014).

Many studies that aim to characterize non-coding transcription have also been carried out on other eukaryotic organisms, including *Saccharomyces cerevisiae*. Even in this primitive eukaryotic unicellular organism, around 85% of the genome is transcribed (David *et al.*, 2006). This trait of transcribing the majority of the genome, a phenomenon often called "permissive transcription", is spread throughout eukaryotes and an increasing number of publications are discussing its potential function(s) (Berretta and Morillon, 2009; Dinger *et al.*, 2009). The identification and characterization of non-coding transcripts has expanded the old definition of a "gene" beyond its protein-coding function. Furthermore, the discovery of both the non-coding genome and the permissive transcriptome has generated heated debates within the scientific community concerning the biological significance and functional pertinence of this DNA and these non-coding RNA, still perceived as dark matter (Clark *et al.*, 2013; Dinger *et al.*, 2009; Mattick, 2003). These debates have challenged Watson and Crick's central dogma of molecular biology by promoting ncRNA at the epicenter of cell processes, as a dynamic motor of biological complexity (Figure 1.9).

The analysis of coding and non-coding RNA expression in the 14 different ENCODE cell lines (Djebali *et al.*, 2012) highlighted the exceptional specificity of the lncRNA for each cell type. While more than 50% of the RNA have a comparable level of expression throughout the 14 cell lines (Figure 1.10), more than 50% of the lncRNA are only detectable in a single cell line. These striking results suggest that lncRNA might have a determining role in cell identity.

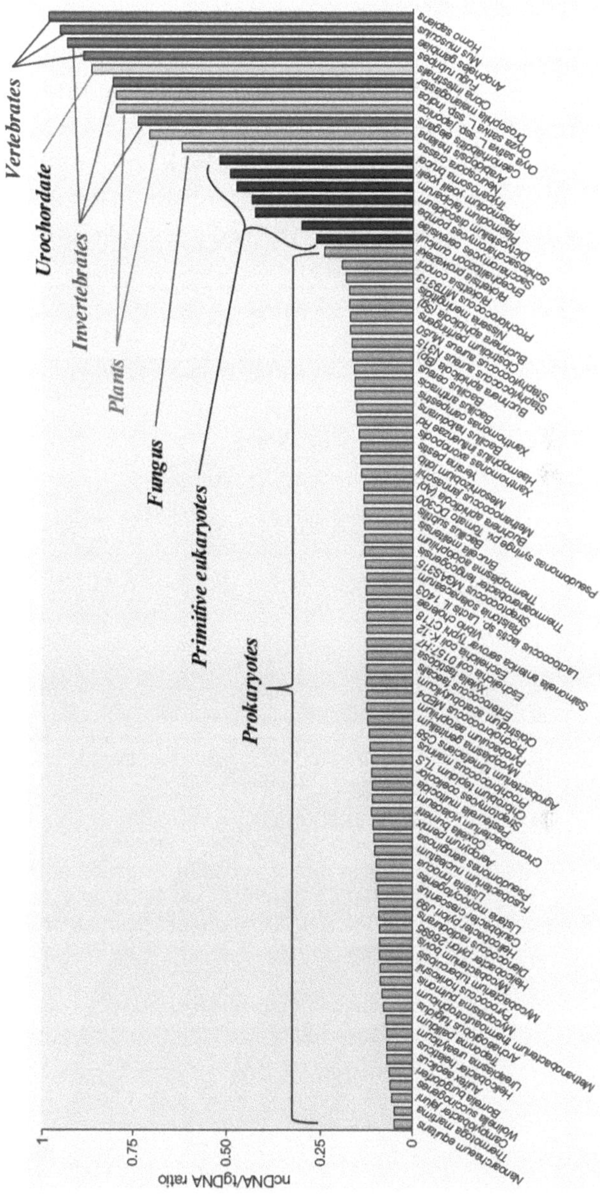

Figure 1.9. *The proportion of the genome that is non-coding increases with the organism's complexity. For a color version of this figure, see www.iste.co.uk/morillon/RNA.zip*

NOTES ON FIGURE 1.9.– Graph representing the proportion of the genome that is non-coding (non-exonic) versus the coding genome (exons) for the sequenced genome of each species (on the abscissa). In light blue: the bacteria; in black: simple eukaryotes (unicellular); in pastel blue: a fungus; in green: plants; in dark blue: invertebrates; in yellow: a urochordate and in red are the vertebrates. Adapted from (Mattick, 2004).

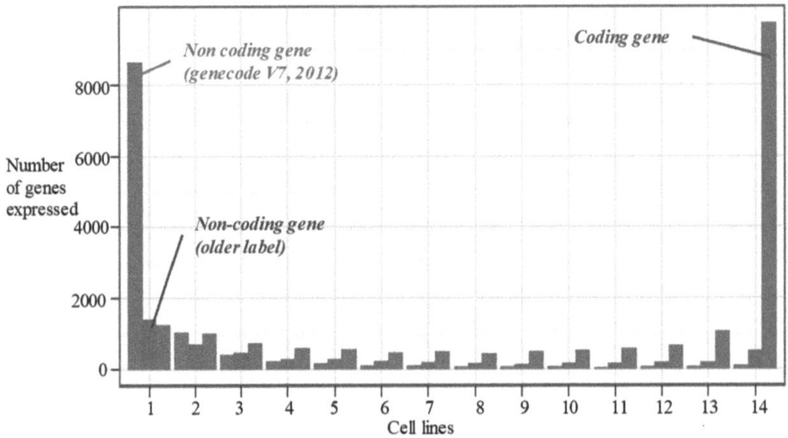

Figure 1.10. *Cell specificity of non-coding gene expression. The number of non-coding genes (coding genes in red; non-coding in green and blue with the 2012 ENCODE label and older label) of which one RNA was detected in one or more cell lines sequenced in the ENCODE project (Djebali et al., 2012). For a color version of this figure, see www.iste.co.uk/morillon/RNA.zip*

2

Definition and Families of Long Non-coding RNA

2.1. The portrait of an ideal suspect in terms of long non-coding RNA

lncRNA has been identified in all species that have been studied at the genomic level including animals, plants, fungi, prokaryotes and even viruses. Genomic level studies continue to add to the catalog of lncRNAs with increasing detail emerging on the specific characteristics of lncRNA as units of transcription. Here, the main characteristics that distinguish ncRNA from mRNA are described (Table 2.1).

2.1.1. *The coding potential of non-coding RNA*

As the acronym suggests, lncRNA genes do not code for proteins. Of course, some lncRNA found in the cytosol are associated with mono- or poly-ribosomal complexes (van Heesch *et al.*, 2014), but this association is not necessarily linked to translation, and has rather been suggested as a degradation and "turnover" factor (Figure 2.1) of these lncRNA (Wery *et al.*, 2016). Some lncRNA include small open reading frames (sORFs) and undergo translation, even though only a minority of these translation events result in stable and operational peptides (Andrews and Rothnagel,

2014; Housman and Ulitsky, 2016). This is the case for DWORF, a muscle-specific lncRNA that codes for an operational peptide of 34 amino acids (Banfai *et al.*, 2012; Derrien *et al.*, 2012). Proteomic studies will no doubt introduce a new "coding" aspect to lncRNA, expanding our concept of the "code" and opening to a potential concept of dual functionality (Ulveling *et al.*, 2011).

	lncRNA	*mRNA*
Transcription		
Mechanism	RNAPII RNAPIII (B2-SINE; NDM29) RNAPIV and V (plants)	RNAPII
Chromatin modification		
H3K4me3	Weak for eRNA, PROMPT	high
H3K4me1	High for eRNA, PROMPT	low
H3K27ac	High	low
H3K79me2	Enriched (bidirectional)	weak
RNA properties		
Runs in 5'	Yes (m7G)	Yes (m7G)
Poly(A) tail	Not always	Always
Length (average)	10 kb (human)	5 kb (human)
Exon-intron composition	Yes	Yes
Splicing	Not always/not effective	Yes
RNA stability	Weak to unstable(eRNA, XUT, CUT, PROMPT)	Variable
Conservation	Weak	High
Coding potential	Yes but small ORFs	Yes
Structure	Multi-modular	No
Subcellular location	Nucleus, cytosol, mitochondria	Cytosol
Expression specificity	Very high	Variable

Table 2.1. *Comparison of lncRNA properties versus those of mRNA*

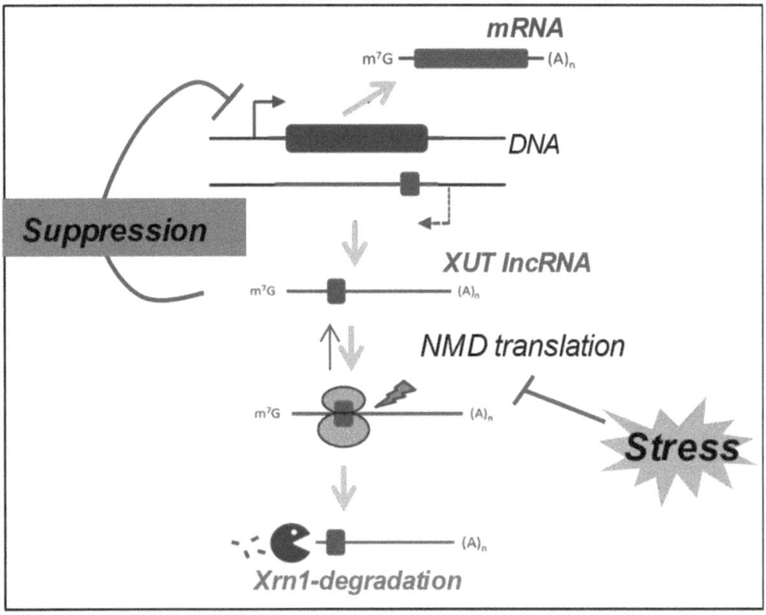

Figure 2.1. *Regulation of XUT stability in the yeast S. cerevisiae. For a color version of this figure, see www.iste.co.uk/morillon/RNA.zip*

NOTES ON FIGURE 2.1.– XUT antisense RNA is recognized by the ribosome which then opens the Nonsense-Mediated-Decay (NMD) degradation channel to activate the Xrn1 exoribonuclease for RNA degradation. Alternatively, if the XUT lncRNA is stabilized by a stress factor that inhibits translation or the NMD, it can play a regulator role (repressor) on gene expression.

2.1.2. *lncRNA transcription and the organization of their transcripts*

The majority of eukaryotic lncRNA are produced by RNA polymerase II (RNAPII), even if there have been exceptions found, such as the B2-SINE RNA introduced by murine thermal shock or MDM29 (Espinoza *et al.*, 2007; Massone *et al.*, 2012), synthesized by RNA polymerase III (RNAPIII). In plants, two specialized RNA polymerases, RNA polymerase IV and polymerase V (RNAPIV and RNAPV), also synthesize certain lncRNA (Ariel *et al.*, 2015) but these remain rare events. Many lncRNA are capped at their 5' end. However, there is some ambiguity around the systematic presence of a cap, especially for those that are highly unstable and weakly abundant. lncRNA can both be or not be polyadenylated at the 3' end and present several isoforms, such as the NEAT1 or MALAT1 transcripts (Djebali *et al.*, 2012; Yang *et al.*, 2011). The lncRNAs that contain a polyadenylation signal seem to have greater stability than those that are poorly polyadenylated or not polyadenylated at all, except for lncRNA that have 3'-specific structures such as MALAT1. It must be noted that transcriptomic studies based on poly(A)+ RNA exclude the characterization of non-polyadenylated transcripts and thus introduce a quantitative bias in the identification of these lncRNA (Eichhorn *et al.*, 2016).

Similarly with PCG, the transcription of many lncRNA genes requires canonical factors such as the pre-initiation complex (PIC), the mediator, the transcription elongation complex as well as specific transcription factors that can in turn define the specific expression of lncRNA in different biological contexts. However, certain distinguishing features in lncRNA promoters have been shown to suggest regulation roles distinct from the transcription of these two gene groups (Alam *et al.*, 2014). For example, the PROMPT promoters lack transcription initiation factors such as TAFI, TAFII, p250 and E2F1, and it appears that transcription bypasses the usual PIC complex (Preker *et al.*, 2011). eRNAs require the Integrator complex for the cleavage of the 3' end of primary transcripts (Lai *et al.*, 2015), while small ncRNA lncRNA precursors are dealt with by specific endonucleases (Dhir *et al.*, 2015; Wilusz *et al.*, 2008). Some unstable lncRNA such as NUT in yeast and CUT are terminated by Nrd1, the transcription termination factor-dependent channel, thus targeting them for rapid degradation by the exosome (Porrua and Libri, 2015).

lncRNA can have a multi-exon composition with splicing signals similar to those found in PCG, and can also produce several different isoforms with distinct functions (Meseure *et al.*, 2016; Spurlock *et al.*, 2015). However, they generally contain fewer exons, which in turn appear longer than those in PCG (Derrien *et al.*, 2012).

2.1.3. *The chromatin signatures of lncRNA genes*

Transcribed by RNA polymerase II, the genomic regions of lncRNA show a chromatin organization similar to PCG but with a few differences (Table 2.1). Overall, the TSS of lncRNA are hypersensitive to DNase I, suggesting a weak nucleosome density as seen in the TSS of PCG. However, lncRNA promoters have inferior levels of histone H3-K4 trimethylation (H3K4me3) compared to PCG, in line with their weak transcription rates. eRNA and PROMPT have high levels of histone H3 K4 monomethylation (H3K4me1) and K27 acetylation (H3K27ac) on their promoters, considered to be specific signatures of unstable transcripts (Marques *et al.*, 2013). Between the TSS and TTS, the majority of lncRNA show a high level of histone H3 K36 trimethylation (H3K36me3) marking the extension phase of transcription, except eRNAs an PROMPT. In yeast, sense-antisense transcription is associated with weak levels of histone H2B ubiquitination, H3K36me3 and H3-K79 trimethylation (H3K79me3), as well as increased levels of H3 acetylation (H3ac), chromatin remodeling enzymes, histone chaperones and histone turnover (Murray *et al.*, 2015). In mice, bidirectional transcription, often associated with development genes and genes involved in transcription regulation, shows high levels of H3K79 dimethylation (H3K79me2) and RNA polymerase II. This signature is characteristic of an early transcription extension phase for a region transcribed in both directions (Lepoivre *et al.*, 2013).

2.1.4. *lncRNA expression: their stability, specificity and abundance*

Several genome-level studies have delved into lncRNA stability and have revealed differences in the functions of lncRNA classes. In mice, measures of lncRNA half-life (t½), carried out by inhibiting transcription in the presence of actinomycin B, showed an average t½ of 4.8 h against 7.7 h in mRNA, suggesting that lncRNA are less stable with an important level of highly unstable members (t½ < 2h) (Clark *et al.*, 2012). The comparison of stability between different classes of lncRNA has revealed that intronic lncRNA or promoter-associated lncRNA are less stable than intergenic, antisense or 3'UTR associated lncRNA. The transcripts containing a single exon are over-represented amongst the most unstable lncRNAs.

In contrast, in human HeLa cells, the same approach for measuring the half-life of lncRNA revealed that antisense lncRNA is more stable than mRNA, while intronic lncRNA remained relatively stable (Ayupe *et al.*, 2015). Circular RNA are examples of very stable lncRNA with a median t½ of 18.8 – 23.7 hours, almost 2.5 times longer than their linear homologs (Enuka *et al.*, 2016).

The RNA degradation channels catalyzed by nuclear and cytoplasmic exosome complexes, cytoplasmic Xrn1 exonuclease and NMD (Nonsense Mediated Decay) degradation, as well as RNAi channels, are known to control the abundance of lncRNA in cells. Circular RNA are intrinsically protected from all degradation channels that depend on exonucleases.

Transcriptomic studies overall have highlighted the spatio-temporal, highly-specific tissue and cell expression profiles of lncRNA compared with PCG. Only a minority are present in all tissue or cell types, such as TUG1 or MALAT1 (Li *et al.*, 2015a; Ward *et al.*, 2015) and the large majority are only expressed in a limited number of different tissues. Curiously, the brain and testicles are sources rich in specific lncRNA expression, supporting the hypothesis that these transcripts are important for the acquisition of specific phenotypic traits (Ward *et al.*, 2015; Washietl *et al.*, 2014). lncRNA expressed in a ubiquitous way are often abundant, whereas specific lncRNA found in one tissue or one cell type appear to be expressed at low levels (Jiang *et al.*, 2016). Disease-associated single nucleotide polymorphisms (SNP) within lncRNA genes and their promoters have been linked with mutations in lncRNA expression, confirming their functional pertinence in certain genetic diseases (Kumar *et al.*, 2013).

The high expression specificity of lncRNA (Figure 1.10) strongly suggests their fundamental regulatory role in different biological contexts, including normal and pathological development.

2.1.5. *The cellular distribution of lncRNA*

Overall, in contrast to mRNA, many lncRNA are found in the nucleus following a focused (foyers) or dispersed (cloud) localization model (Cabili *et al.*, 2015). There are, however, many examples of those present in the nucleus and cytosol such as TUG1 or HOTAIR, or exclusively in the cytosol such as DANCR (Djebali *et al.*, 2012). Multiple determinants, such as a specific RNA template (Zhang *et al.*, 2014) or the formation of particular RNA-protein complexes, can dictate their subcellular location and define their potential function (Chen, 2016). Remarkably, environmental changes or infections can induce the delocalization (or active traffic) of lncRNA from one cell compartment to another, as is the case for lncRNA induced through stress (Giannakakis *et al.*,

2015). Thus, HuR and GRSF1 modulate the nuclear exportation and mitochondrial location of the RMRP protein (Noh *et al.*, 2016).

2.1.6. *The structure of lncRNA*

RNA is a very flexible and dynamic molecule that forms complex secondary structures. The structure of lncRNA defines its cellular degradation and no doubt its functional versatility, allowing it to set up its subcellular location and form interactions with protein complexes (Lu and Chang, 2016). An increasing number of examples show that the secondary structure of lncRNA constitutes the main primary functional unit, thus bypassing the absence of evolutionary constraints (no code for proteins) and their poor sequence conservation between living species (Johnsson *et al.*, 2014). HOTAIR is the most studied example, characterized only in mammals, and sharing only 58% homology between mice and men (Bhan and Mandal, 2015; He *et al.*, 2011), but with a conserved function of epigenetic regulation of chromatin domains. The analysis of covariance of HOTAIR across 33 mammal genomes revealed a significant number of covariant base pairs, thus HOTAIR has a similar structure regardless of the modified sequence (Somarowthu *et al.*, 2015). On the other hand, poor sequence conservation, which allows for structural changes, can lead to the rapid acquisition of new functions and to the specialization of regulatory circuits. Such is the case for lncRNA associated with regions with accelerated mutation rates in humans (HAR, Human Accelerated Region), also expressed in the developing neocortex of primates, but where the capacity to form a stable structure in a clover shape has only ever been seen in humans (Beniaminov *et al.*, 2008; Pollard *et al.*, 2006b). It has been suggested that these lncRNA may have a particular specificity in humans, with a role in the unique capacities of the human brain.

Many structural prediction tools, such as R-fold, have been developed to provide indications of function (Kiryu *et al.*, 2008). Structural analysis of RNA has increased our understanding of the mechanical aspects of lncRNA activity. However, X-ray crystallography, Nuclear Magnetic Resonance (NMR) and cryo-electronic microscopy (CryoEM) require stable, nearly static molecules and are not adapted for highly dynamic and flexible RNA. Very recently, new technologies based on high throughput sequencing (of the SHAPE variety) have enabled an *in vitro* and *in vivo* view of lncRNA structure (Lu and Chang, 2016).

2.2. Classification of lncRNA

Progresses in sequencing technology have given rise to a plethora of new transcripts requiring the establishment of a universal standardized system for the classification of lncRNA and their functional labeling. This standardization is only in its early stages and different classifications based on length, transcription properties, position in terms of known genomic labels, regulator elements and function have been proposed. Table 2.2 shows a non-exhaustive catalog of eukaryotic lncRNA based on these different classifications.

2.2.1. *Based on size*

Conventionally, a length of more than 200 nucleotides is the limit by which long ncRNA (lncRNA) and short RNA (sRNA) are discriminated. However, lncRNA vary considerably in size and those greater than 10,000 bases (kb) belong to the group of very long intergenic RNA (vlinc- or macro-lncRNA). These transcripts have particular characteristics that distinguish them from other lncRNA: they are either badly or not at all spliced, weakly polyadenylated at the 3' end, and are produced from specific genomic loci (Table 2.2). The majority of vlincRNA are located close to or within PCG promoters on the same or the opposite strand, and work in *cis* as positive transcription regulators of the nearby genes. Interestingly, certain vlincRNA promoters carry Long Terminal Repeats (LTR) that are highly regulated by three main transcription factors associated with pluripotency, suggesting a possible role in early embryonic development (St Laurent *et al.*, 2013). Others are specifically induced during senescence and are necessary for the maintenance of senescent characteristics that, in turn, control the transcriptional response to environmental changes (Lazorthes *et al.*, 2015). Macro-lncRNA are often antisense to PCG and are produced from parental footprint regions. Macro-lncRNA can either switch off nearby genes by setting off chromatin modification during their transcription or through a transcriptional interference mechanism (Guenzl and Barlow, 2012).

2.2.2. *Based on position relative to gene promoters*

The classification of lncRNA based on their position relative to gene promoters is currently used for the GENCODE/Ensemble portal but also by consortia and laboratories for newly assembled transcripts. Initially, transcripts were either classed as intergenic or intragenic (Figure 2.2). Long intergenic non-coding RNA (lincRNA) do not intersect with coding genes labels (Guenzl and Barlow, 2012). lincRNA are shorter than PCG, transcribed by RNA polymerase II, contain a 5’ cap, are polyadenylated in 3’ and spliced. Although several highly conserved lincRNA exist, the majority of them have poor sequence conservation and contain 5’ patches nested within exons (Hezroni *et al.*, 2015). It is admitted that highly conserved lincRNA contribute to biological processes common across many lineages, such as embryonic development (Necsulea *et al.*, 2014), whereas others ensure phenotype and functional variation. Many, if not nearly all, lincRNA are found in the nucleus where they carry out their regulatory functions. For example, lincRNA-p21 is induced by p53 following damage to DNA (Huarte *et al.*, 2010). It physically connects to the hnRNP-K nuclear factor on promoters specific to p52-dependent transcriptional responses.

lincRNA that overlap PCG labels can be classed as sense or antisense, bidirectional and/or intronic lncRNA.

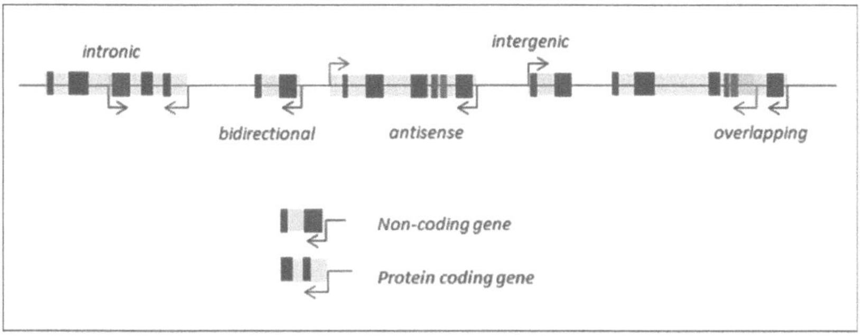

Figure 2.2. *lncRNA families based on their genomic position. For a color version of this figure, see www.iste.co.uk/morillon/RNA.zip*

NOTES ON FIGURE 2.2– Annotation schematic of non-coding transcripts based on their genomic position relative to coding genes (blue box, coding exon; red box, non-coding exon).

2.2.2.1. *Antisense lncRNA*

Antisense lncRNA (Figure 2.2) of coding genes (asRNA or ancRNA) were first discovered following single gene studies, but recent developments in RNA sequencing (RNA-seq) have allowed for their systematic identification as fundamental components of eukaryotic transcriptomes (Goodman *et al.*, 2013; Kapranov *et al.*, 2005; Wood *et al.*, 2013). This group encompasses so-called natural antisense transcripts (NATs) that are in turn subdivided into cis-NAT that affect the expression of corresponding sense transcripts, and trans-NATs that regulate the expression of non-paired genes from other genomic sites (Magistri *et al.*, 2012; Su *et al.*, 2010; Yuan *et al.*, 2015).

A recent study highlighted the greater expression specificity and the increased stability of asRNA compared with lincRNA and intragenic sense lncRNA (Ayupe *et al.*, 2015). Due to their sequence complementarity with mRNA or sense pre-mRNA, asRNA can work by forming RNA-RNA pairs, ensuring the specific targeting of their potential regulatory activity, and no doubt increased stability too (Figure 2.3). This is the case for BACE1-AS, strongly expressed in Alzheimer's disease cellular models, which stabilizes BACE1 mRNA and translates into both an increased expression of beta-secretase coded by BACE1 and the accumulation of beta-amyloid peptides in the brain (Faghihi *et al.*, 2008).

It has also been shown that antisense transcription across intronic regions regulates the local organization of chromatin, thus affecting co-transcriptional splicing of sense pre-mRNA (Gonzalez *et al.*, 2015). In addition, certain NAT contain the interspersed and inversed B2 repeat element (SINEB2) such as AS-Uchl1 (Carrieri *et al.*, 2012). These NAT, called SINE-Ups, are capable of stimulating the translation of their sense mRNA thanks to an overlapping 5' sequence complementary to the coding gene. Recently, SINEUPs have been suggested for biotechnological and therapeutic application in the treatment of haplo-insufficiency (Indrieri *et al.*, 2016; Zucchelli *et al.*, 2015). Despite the poor evolutionary conservation of antisense transcription, certain lncRNA subgroups, such as vlincRNA and macro-lncRNA associated with senescence in mammals, or XUT in yeast, are mainly defined as antisense, suggesting regulation pathways dedicated to the biogenesis of these transcripts, such as the control of cell homeostasis, the stress response and associated diseases (Wood *et al.*, 2013).

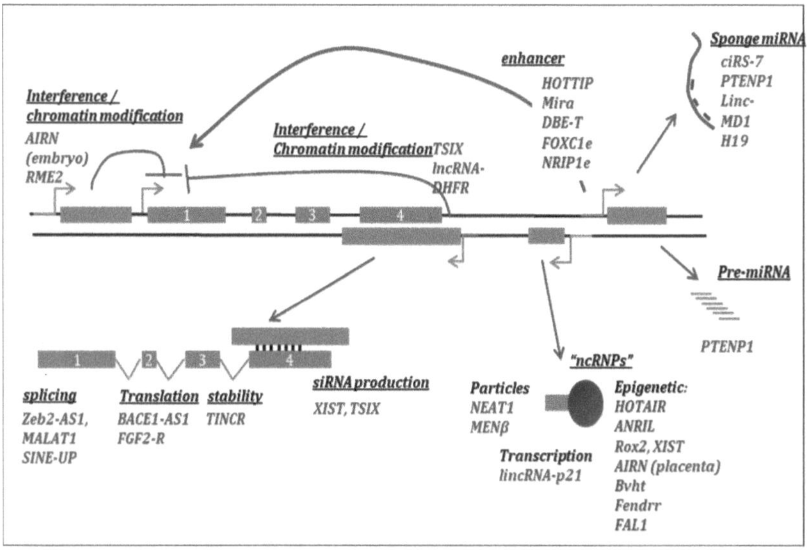

Figure 2.3. *Long non-coding RNA in the regulation of gene expression. For a color version of this figure, see www.iste.co.uk/morillon/RNA.zip*

NOTES ON FIGURE 2.3– The different forms of regulation that have been suggested for the best characterized mammalian examples in the literature: from transcriptional interference to the regulation of splicing, the translation of ncRNPs (non-coding ribonucleoparticles) and precursor miRNA (pre-miRNA) formation. Non-coding genes are in red and coding genes in blue. The examples described in the literature are shown in light blue.

2.2.2.2. *Bidirectional lncRNA*

The discovery of bidirectional transcription as an intrinsic characteristic of eukaryote transcription machinery gave rise to the identification of bidirectional lncRNA (Kapranov *et al.*, 2005; Scruggs *et al.*, 2015; Seila *et al.*, 2008; Wei *et al.*, 2011; Xu *et al.*, 2009). These transcripts either do not or only partially cover the 5' regions of paired PCG, and several nomenclatures have been suggested for their classification:

paRNA, LUAT and uaRNA (Flynn *et al.*, 2011a; Hamazaki *et al.*, 2015; Hung *et al.*, 2011; Lepoivre *et al.*, 2013; Uesaka *et al.*, 2014). Currently, the number of bidirectional lncRNA is grossly underestimated not only due to the inexact labeling of TSS and promoters in the genome, but also because of their highly unstable nature and the difficulties linked with their detection. Genomic studies have revealed that bidirectional promoters show unusual sequences and epigenetic characteristics. In addition, they can be found close to genes involved in particular biological processes such as development transcription factors or cell cycle regulation (Hu *et al.*, 2014a; Hung *et al.*, 2011; Lepoivre *et al.*, 2013; Sigova *et al.*, 2013; Uesaka *et al.*, 2014). It appears that the misbalance in bidirectional transcription constitutes a final indigenous control mechanism that is particularly effective during the activation or suppression of a gene (Figure 2.3) (Kambara *et al.*, 2014; Morris *et al.*, 2008).

2.2.2.3. *Intronic and circular lncRNA*

Intronic lncRNA are limited to PCG introns and can either be single transcripts produced from an autonomous transcription, or a sub-product of the pre-mRNA process during splicing. Examples of intronic transcripts derived from pre-mRNA are circular intronic RNA (ciRNA) produced from intron lassos (lariat) that escaped dissociation (debranching) before their degradation (Zhang *et al.*, 2013), but also sno-lncRNA produced from introns containing two incorporated snoRNA genes (Yin *et al.*, 2012). The latter appear to positively regulate the transcription of coding genes and their splicing by accumulating near the transcription site.

Intronic transcripts expressed independently from host gene transcription are considered as the most widespread of intronic lncRNA (St Laurent *et al.*, 2012).

2.2.2.4. lncARN *that overlap with messenger RNA*

The labeling of lncRNA sense transcripts, overlapping or enveloping entire exons or coding genes, includes a portion of the homonyms adopted by GENCODE under the "GENENAME-OT" label. For example, in its intron, SOX2-OT accommodates a gene coding for one of the main pluripotency regulators, the SOX2 gene. SOX2-OT is expressed in a dynamic manner and is spliced alternatively not only during differentiation, but also in cancer cells where it has been suggested that it regulates the expression of SOX2 (Shahryari *et al.*, 2015).

Additionally, the lncRNA that overlap with coding genes are able to form circular lncRNA (circRNA, Figure 2.3) following non-canonical splicing (Hansen *et al.*, 2013; Memczak *et al.*, 2013). Certain sequence characteristics such as the presence of repeat elements in introns could be key to the activation of this non-canonical splicing and the generation of a circular RNA molecule (Kramer *et al.*, 2015). For example, it has been suggested that Alu elements within introns participate in the circularization of RNA via RNA-RNA pairing (Hadjiargyrou and Delihas, 2013). Remarkably, these events appear to be specific to certain tissues or cells, or limited to certain development stages and pathological contexts (Peng *et al.*, 2015; Rybak-Wolf *et al.*, 2015). More generally, circRNA operate in the cytosol as miRNA sponges, as seen with CDR1as/ciRS-7 which binds a multitude of miR-7 and prevents them from finding their initial mRNA target (Hansen *et al.*, 2013; Memczak *et al.*, 2013).

Certain circRNA called exon-intron circRNA (EIciRNA) contain an additional non-spliced intron that ensures that the circRNA are retained in the nucleus to interact with U1 snRNP or to promote the transcription of their parental genes (Li *et al.*, 2015e). The most remarkable property of circRNA is their stability which makes them eligible as powerful diagnostic markers and therapeutic agents (Li *et al.*, 2015c).

2.2.3. *Based on their position relative to DNA regulator elements*

2.2.3.1. *lncRNA derived from pseudogenes*

In addition to PCGs, mammalian genomes contain dozens of thousands of pseudogenes that are the genomic remains of ancient coding genes that have lost their coding potential through the course of evolution. Many of them are transcribed in lncRNA sense and antisense directions. Given the sequence similarity with existing parental genes, lncRNA derived from pseudogenes can thus regulate the expression of PCGs through intermediate RNA-RNA hybrids by acting as miRNA sponges, and producing endogenous siRNA (Figure 2.3) or integrating directly with mRNA (Milligan and Lipovich, 2014; Zheng *et al.*, 2007). PTENP1, a lncRNA pseudogene derived from the PTEN tumor suppressor gene, was amongst the first described as an miRNA sponge with a role in tumor progression (Poliseno *et al.*, 2010).

2.2.3.2. *lncRNA derived from ultra-conserved regions*

Ultra-conserved regions (UCR) are segments of the genome (≥ 200 base pairs) that show 100% sequence conservation between humans, mice and rats. The human genome contains 481 UCR of which 39% are intergenic, 43% intronic and 15% are exonic (Bejerano *et al.*, 2004). These regions are largely transcribed in T-UCR lncRNA (Mestdagh *et al.*, 2010; Watters *et al.*, 2013). The expression of T-UCR is induced through stress linked to tumor development such as retinoid treatment or hypoxia. They are expressed abnormally in different cancers and some are associated with a poor prognosis (Fassan *et al.*, 2014; Ferdin *et al.*, 2013; Watters *et al.*, 2013). Given their high expression specificity,

T-UCR have been put forward as molecular markers for cancer diagnostics and prognosis (Scaruffi *et al.*, 2009). However, the function of T-UCR is still poorly understood. Evf2 (or Dlx6as) is an example of a T-UCR with an "attractant" role. It interacts with the DLX1 transcription activator, increasing its association with the SWI/SNF-like complex linked to Brahma 1 (BRG1), inhibiting its ATPase activity and therefore its chromatin remodeling activity (Cajigas *et al.*, 2015; Feng *et al.*, 2006).

2.2.3.3. *Repeat-containing lncRNA*

Telomeres are protector nucleo-protein structures located at chromosome extremities. They are transcribed into telomere repeat-containing non-coding RNA, TERRA. This family of transcripts is generated from both the Watson and Crick strands in a cell cycle-dependent manner (Feuerhahn *et al.*, 2010; Porro *et al.*, 2010). It has been suggested that the formation of RNA-DNA hybrids containing TERRA, at the extremities of chromosomes, regulates the size or the erosion of telomeres and thus controls senescence (Feuerhahn *et al.*, 2010; Porro *et al.*, 2010). The sub-telomere regions are also actively transcribed (Broadbent *et al.*, 2015; Greenwood and Cooper, 2012; Trofimova *et al.*, 2015). In budding yeast, the heterogeneous population of lncRNA, called subTERRA, accumulates in a transitory manner in the G2/M and late G1 phases of the cell cycle (Kwapisz *et al.*, 2015). The exact function of subTERRA is not clear, although it seems to have a regulatory role in the homeostasis of telomeres.

Recent discoveries have revealed that centromere repeats are also actively transcribed into lncRNA during the late stages of mitosis and the beginning of the cell cycle G1 phase (Blower, 2016; Chan *et al.*, 2012; Quenet and Dalal, 2014; Rosic *et al.*, 2014; Wong *et al.*, 2007). These lncRNA physically interact with different components of specific centromere nucleoproteins such as CENP-A/C and HJURP, and seem to be important in the maintenance of centromere integrity.

Finally, it has been shown that certain ribosomal DNA (rDNA) regions are transcribed by RNA polymerase II, antisense to the rRNA genes, producing a heterogeneous population of lncRNA called PAPAS (promoter and antisense of rRNA). Their expression is induced in quiescent cells and initiates the recruitment of the H4K20 methyltransferase Suv4-20h2 histone on the rDNA genes thus inducing transcriptional silence (Bierhoff *et al.*, 2014). PAPAS also enable the formation of heterochromatin and stop the expression of certain genes in limited cells for their growth.

2.2.3.4. *Gene enhancer and promoter region lncRNA*

Promoters and "enhancers" constitute cis-regulator elements fundamental to the control of gene expression, providing platforms for the recruitment of transcription factors and for the establishment of a specific chromatin organization. Many, if not all, functional enhancers are transcribed permissively in both directions – sense and antisense – and are called eRNA. Transcribed enhancer regions have signatures that are specific at the histone level, distinguishing them from other transcription units. Such signatures include a higher level of the H3-K27ac and H3K4me1 histones compared with other lncRNA and PCGs. The end of the eRNA depends on the Integrator complex, maturing the transcript at the 3' extremity without recruiting the polyadenylation complex, thus resulting in highly unstable RNA. Their expression is specific to the cell type, the tissue type or the development stage and can be activated by external or internal stimuli.

The transcription of enhancers seems to mark a positive transcription activity for the expression of associated genes (Wang *et al.*, 2011a). However, the role of eRNA remains controversial and only a few examples have been clearly described such as FOXC1e or NRIP1e (Figure 2.3) (Li *et al.*,

2013b). It has been suggested that these eRNA control the promoter's chromatin environment, the formation of a DNA loop between the promoter-enhancer, the charging and stopping of RNA polymerase II and thus the increase in transcription factor concentration at the local level. All these contribute to the strong activation of nearby gene transcription (Li *et al.*, 2016).

Promoter-associated lncRNA (PALRs) are transcribed in sense and antisense directions at promoter region level and can partially overlap with the 5' extremity of the gene (Kapranov *et al.*, 2007). This class of transcripts includes highly unstable PROMPTs and antisense RNA located upstream of the promoters (uaRNA) that are easily detectable in a context where the nuclear exosome has been inactivated (Flynn *et al.*, 2011a; Preker *et al.*, 2011 ; Preker *et al.*, 2008). The polyadenylation-dependent degradation of PROMPT also seems to be involved in controlling the orientation of transcription of bidirectional promoters (Ntini *et al.*, 2013). The presence of an intron competent for splicing in uaRNA could therefore facilitate the formation of a DNA loop by placing termination factors beside a bidirectional promoter thus ensuring the coding gene directionality of RNA polymerase II (Agarwal and Ansari, 2016). In contrast, certain PARL have been shown to negatively regulate the transcription of nearby genes. For example, the PALR of the CCND1 gene promoter suppresses transcription by locally inhibiting the acetyltransferase CBP/p300 histone activity downstream of the target gene, cycline D1 (Song *et al.*, 2012; Wang *et al.*, 2008).

2.2.4. *Based on their biogenesis and degradation*

In budding yeast, given that many lncRNA are highly unstable or "cryptic", the classification currently used is based on their degradation or biogenesis characteristics. However, certain so-called stable unlabeled transcripts

(SUT) have been identified with a wild type genetic pool (Xu *et al.*, 2009). Others are only detectable in specific stress conditions or in lines with an RNA degradation mutation. The latter are divided into three classes: cryptic unstable transcripts (CUT) that are sensitive to the nuclear RNA degradation pathway (Neil *et al.*, 2009; Xu *et al.*, 2009), transcripts whose ending is dependent on Nrd1 (NUT) (Schulz *et al.*, 2013, and unstable transcripts sensitive to the Xrn1-dependent cytoplasmic degradation pathway (XUT) (Berretta *et al.*, 2008; Van Dijk *et al.*, 2011). While the majority of XUT are transcribed in the antisense to PCGs, CUT are often bidirectional but can also be antisense if the upstream coding gene is sufficiently close to the bidirectional promoter (Camblong *et al.*, 2009). Beyond each class definition, there is considerable overlap between CUT and NUT, but also between XUT and SUT (Fox *et al.*, 2015; Wery *et al.*, 2016). Some CUT escape nuclear degradation and are exported into the cytoplasm where they are taken up by Xrn1 (Van Dijk *et al.*, 2011). Another sub-class of CUT is induced during meiosis (MUT) and is degraded by the Rrp6 nuclear exosome (Lardenois *et al.*, 2010). The key difference between the CUT, XUT and SUT is determined by distinct subcellular positions (Figure 2.4). CUT are transcribed and degraded in the nucleus whereas SUT and XUT are exported towards the cytoplasn where many XUT are degraded by Xrn1 unless they pair up with complementary mRNA (Wery *et al.*, 2016). In this case, they can be protected and eventually translated into peptides to give rise to new functional molecules (de Andres-Pablo *et al.*, 2016).

In other eukaryotes, certain lncRNA are also very unstable: previously mentioned PROMPT and eRNA could be considered as human analogues of CUT since they are heavily stabilized during the depletion of human nuclear exosome (Preker *et al.*, 2011). Until now, the existence of XUTs in mammals has not been uncovered, but it is highly likely that an unidentified group of lncRNA would be an appropriate substrate for this exonuclease, similarly to yeast.

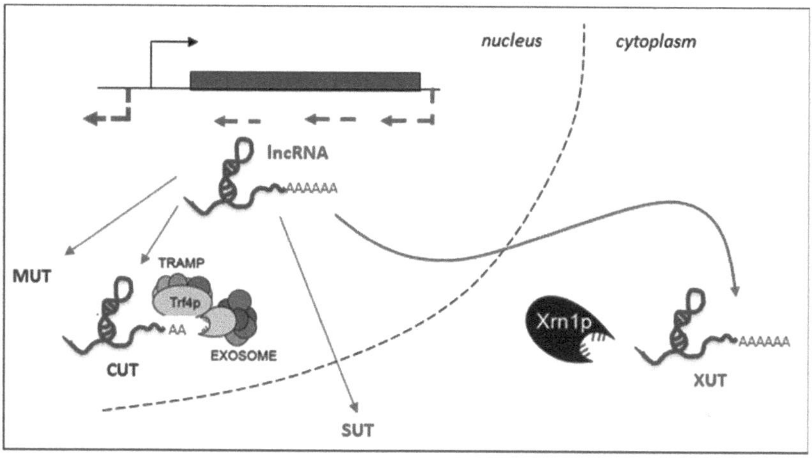

Figure 2.4. *lncRNA in* S. cerevisiae. *For a color version of this figure, see www.iste.co.uk/morillon/RNA.zip*

NOTES ON FIGURE 2.4.– The different families of lncRNA in cerevisiae. Xrn1-sensitive Unstable Transcripts (XUT) are degraded by Xrn1p in the cytoplasm. Cryptic Unstable Transcripts (CUT) re degraded by the exosome in the nucleus. Meiotic Uncharacterized Transcripts (MUT) are expressed during meiosis and Stable Unannotated Transcripts (SUT) are stable. MUT and CUT are divergent of genes while SUT and XUT are antisense.

2.2.5. *Based on their distribution in the cell*

The lncRNA subcellular location can provide a fundamental indication of its function. lncRNA can be exclusively cytosolic (DANCR or OIP5-AS1) or nuclear (NEAT1) or have dual locations (HOTAIR) (Ayupe *et al.*, 2015). Several lncRNA sub-groups with precise subcellular localizations have been defined, such as chromatin associated lncRNA, or CAR (Mondal *et al.*, 2010; Werner and

Ruthenburg, 2015). They may have a role in the assembly of sub-nuclear domains or RNP complexes, activating factors involved in chromatin modifications (Singh and Prasanth, 2013). GAA repeat-containing RNA (GRC-RNA), which represent a sub-class of nuclear lncRNA, form foci in the nucleus during interphase in mammals, where they form part of the nuclear matrix. They appear to play a role in nuclear organization by reassembling various proteins associated with the nuclear matrix (Zheng *et al.*, 2010).

The mitochondrial genome is also transcribed in mitochondrial ncRNA, called ncmtRNA (Anandakumar *et al.*, 2015; Burzio *et al.*, 2009; Rackham *et al.*, 2011). Their biogenesis depends on the mitochondrial proteins coded by the nucleus. After synthesis, certain ncmtRNA are exported from the mitochondria to the nucleus (Landerer *et al.*, 2011).

2.2.6. *The labeling of lncRNA: facing the challenge*

The discovery of new lncRNA is now a daily event and their correct labeling and classification have become a necessity. On top of catchy nicknames, various lncRNA classifications that are based on specific transcript properties, origins or possible functions have been suggested through oral and written communications. Database sites, such as RNAcentral and other consortia, have the difficult task of integrating unambiguous labels of all transcripts and genes, including numeric identifiers as well as unique transcript names. John Mattick and John Rinn have recently suggested several rules for the labeling of lncRNA. Notably, it has been recommended that intergenic lncRNA be named "LINC-X" in which X represents a number, whereas all intragenic lncRNA are identified by their corresponding "GENENAME" with an "AS" prefix for the antisense, "BI" for bidirectional, "OT-" for those overlapping with sense genes and "INT-" for intronic transcripts, in order to provide information based on position (Mattick and Rinn, 2015). By respecting this guideline, OT-SOX2-1 would correspond to the first isoform of the SOX2-OT1 lncRNA, overlapping in the sense direction with the SOX2 gene, whereas HOTAIR would be named AS-HOXC11-1 to designate the largest antisense lncRNA of the HOXC11 gene. However, the original lncRNA nicknames attributed experimentally must be preserved to maintain a historical record and for the follow-up of studies on them.

Identification, labeling and classification are the first steps towards understanding the biology of lncRNA and the non-coding genome. This work is still in its early phases and needs innovative ideas and methods, as well as the development of new, more precise technology and bioinformatics tools dedicated to the discovery and assembly of transcriptomes. The beginning of the 21st Century, in particular, was marked by the emergence of new genomic and systems biology technologies. These new approaches will allow researchers to build a complete picture of the regulatory circuits integrating both coding and non-coding transcription, thus deciphering a little more from the enormous source of information represented by the genome.

lncRNA categories	Abbreviation	Examples	Refs.
Nomenclature based on size			
Long non-coding RNA Large non-coding RNA	lncRNA		(Djebali *et al.*, 2012), (Derrien *et al.*, 2012)
Very long lncRNA	vlincRNA		(St Laurent *et al.*, 2016)
Macro lncRNA		*XIST, Airn, Kcnq1ot1*	(Guenzl and Barlow, 2012)

Nomenclature based on position relative to the promoter and coding genes				
Intergenic lncRNA		lincRNA	*MALAT1, NEAT1, GAS5, CYRANO, Frigidair, lincRNA-COX2, XACT*	(Clemson *et al.*, 2009; Guttman *et al.*, 2009; Kino *et al.*, 2010; Lin *et al.*, 2007; Nam and Bartel, 2012; Ulitsky *et al.*, 2011; Vallot *et al.*, 2013)
Antisense lncRNA				
	Natural antisense	NAT *cis*-NAT *trans*-NAT	*ZEB2NAT* *BACE1-AS*	(Beltran *et al.*, 2008; Faghihi *et al.*, 2008; Nam and Bartel, 2012)
	Antisense lncRNA	asRNA ancRNA		
	Containing the SINE B element	SINE-UP	*AS-Uchl1*	(Zucchelli *et al.*, 2015)
Bidirectional lncRNA	Long upstream antisense transcript	LUAT		(Lepoivre *et al.*, 2013)
	Upstream antisense RNA	uaRNA		(Flynn *et al.*, 2011a)

Intronic lncRNA				
	Totally intronic	TIN		(Nakaya *et al.*, 2007)
	Circular	ciRNA circRNA	*ci-ankrd52, CDR1as/ciRS-7, cANRIL*	(Jeck *et al.*, 2013; Kramer *et al.*, 2015; Memczak *et al.*, 2013; Salzman *et al.*, 2012; Zhang *et al.*, 2013)
	Stable	sisRNA	*sisR-1*	(Gardner *et al.*, 2012; Pek *et al.*, 2015; Talhouarne and Gall, 2014)
Overlapping in tandem		Sense ncRNA	*HLXB9-lncRNA, SOX2-OT*	(Kapranov *et al.*, 2005; Shahryari *et al.*, 2015)
Classification based on position in a DNA regulator element or in a chromatin region				
Pseudogenes			*PTENP1, Lethe*	(An *et al.*, 2017; Poliseno *et al.*, 2010; Rapicavoli *et al.*, 2013)

Telomeres and subtelomeres			*TERRA*	(Azzalin *et al.*, 2007)
			subTERRA	(Broadbent *et al.*, 2015; Greenwood and Cooper, 2012; Kwapisz M, 2015; Trofimova *et al.*, 2015; Vembar *et al.*, 2014),
Centromeres			*centromeric alpha-satellite RNA*	(Blower, 2016; Quenet and Dalal, 2014; Rosic *et al.*, 2014; Wong *et al.*, 2007)
Ultra-conserved regions		T-UCR	*Uc.283+A* *Evf2*	(Feng *et al.*, 2006; Ferdin *et al.*, 2013; Liz *et al.*, 2014; Watters *et al.*, 2013)
lncRNA associated with the Promoter		pancRNA PALRs	*CCND1-lncRNA*	(Liz *et al.*, 2014)
	Upstream of Promoters	PROMPT		(Preker *et al.*, 2008)
	Antisense of promoters	uaRNA*		(Flynn *et al.*, 2011a)
Associated with Enhancer regions		eRNA	*IL1β-eRNA, FOXC1e*	(Li *et al.*, 2013b; NE *et al.*, 2014)

Associated with 3'UTR		uaRNA*		(Mercer *et al.*, 2011)
Classification based on their biogenesis				
Stable (yeast)		SUT		(Xu *et al.*, 2011)
Unstable (yeast)		CUT	*PHO84 CUT, PROMPT, eRNAs*	(Camblong *et al.*, 2009; Neil *et al.*, 2009; Xu *et al.*, 2011)
	Unstable but expressed in meiosis	MUT		(Lardenois *et al.*, 2010)
Dependent on Nrd1		NUT		(Schulz *et al.*, 2013)
Sensitive to Xrn1		XUT	*RTL, XUT1678 (ARG1-AS)*	(Berretta *et al.*, 2008; Van Dijk *et al.*, 2011)
Classification based on location in the cell				
Nuclear lncRNA			*NEAT1*	(Ayupe *et al.*, 2015)
	Containing GAA repeats	GRC-RNA		(Zheng *et al.*, 2010)
	Enriched or associated with chromatin	cheRNA, CAR		(Mondal *et al.*, 2010; Werner and Ruthenburg, 2015)

Mitochondrial lncRNA		ncmtRNA	*ASncmtRNA-2*	(Anandakumar *et al.*, 2015; Bianchessi *et al.*, 2015)
Classification based on biological role				
Scaffolding			*HOTAIR, LINP1, NORAD*	(Lee *et al.*, 2016; Yoon *et al.*, 2013; Zhang *et al.*, 2016)
	Architects	arcRNA	*NEAT1*	(Chujo *et al.*, 2016)
Guides			*MEG3, Khps1*	(Mondal *et al.*, 2015; Postepska-Igielska *et al.*, 2015)
Ribo-activators		ncRNA	*SRA, Lnc-DC, NeST*	(Gomez *et al.*, 2013; Orom *et al.*, 2010; Wang *et al.*, 2014; Yao *et al.*, 2010)
		eRNA	*FOXC1e*	(Li *et al.*, 2016)
Ribo-suppressor or lure			*GAS5, CCND1-lncRNA, PANDA, Lethe*	(Hung *et al.*, 2011; Kino *et al.*, 2010; Pek *et al.*, 2015; Song *et al.*, 2012)

Competitor (titration)	ceRNA	*PTENP1, HULC, CDR1as/ciRS-7*	(An *et al.*, 2017; Memczak *et al.*, 2013; Tay *et al.*, 2014)
Precursors	endo-si-lncRNA pi-lncRNA mi-lncRNA	*H19 (miR-675)* *MALAT1 (mascRNA)* *P5CDH-SRO5*	(Carlile *et al.*, 2009; Giovarelli *et al.*, 2014; Ha *et al.*, 2014; Wilusz *et al.*, 2008)
Classification based on biological process			
Induced by hypoxia	HINCUT		(Choudhry *et al.*, 2016; Ferdin *et al.*, 2013)
Associated with senescence	SAL	*vlincRNAs, SALNR*	(Lazorthes *et al.*, 2015; Wu *et al.*, 2015)
Induced through stress	si-lncRNA		(Giannakakis *et al.*, 2015)
Associated with prostate cancer	PCATs	*PCA3, PCAT1*	(Prensner *et al.*, 2011)

Table 2.2. *Classification of lncRNA (adapted from (St Laurent et al., 2015)). In the literature, the term "uaARN" is used for two distinct groups of transcripts: upstream antisense RNA and UTR-associated A*

3

Biological Functions of Long Non-coding RNA

3.1. Non-coding RNA: rejects or functional elements of genomes?

The controversy around the biological and functional pertinence of lncRNA is essentially based on two observations.

– The first concerns the origins, the infinite diversity of lncRNA across species and most notably the absence of evolutionary conservation of primary sequences, indicating weak selection pressure and the consequent relative functionality.

– The second argument concerns the very rare identification of lncRNA during genetic screens in the different model organisms studied, again potentially reflecting their limited role in the fundamental cell pathways explored by these studies.

However, despite these debates, the proof that certain lncRNA clearly play first-line roles in development, pluripotency, dosage compensation, the establishment of chromatin structure, genome maintenance, and as tumor suppressors or oncogenes is not contested. It is now admitted that although they may not all have a direct role, the function of lncRNA is not limited to their primary sequence but probably to complex secondary structures, providing an explanation of why genetic screens targeting exons or base inversions in the genome (often exclusively coding) would be insufficient to dramatically affect their functions.

3.1.1. *Where do non-coding RNA come from?*

Hypotheses on the origins of non-coding genes are varied and several mechanisms have been suggested (Figure 3.1): existing DNA or RNA sequence-based genome duplications, the mutation of PCGs through the loss of their protein-coding potential, the random integration or excision of transposable elements or non-coding DNA (Marques and Ponting, 2014). For example, pseudogenes are an example of PCGs in which one of the gene copies (having duplicated or multiplied) accumulated, over time and through evolution, mutations in its original open reading frame thus removing its translation potential.

Once transcribed, pseudogenes often produce lncRNA, as seen with PTENP1. The pseudogenization of a PCG, due to mutations that are deleterious for translation, can equally produce lncRNA that do not have an obvious "homolog" coding for a protein. For example, it has been suggested that the XIST lncRNA is derived from the ancestral Lnx3 gene that acquired several mutations that, through the rapid evolution of placental mammals, therefore shifted the reading frame (Duret *et al.*, 2006).

The integration or excision of transposable elements (TE) in non-coding genes is another frequent mechanism used to explain the origins of lncRNA. In humans, TE constitute a large portion of the genome, potentially representing up to

40-45% of it (Lander *et al.*, 2001a). The majority of them are genomic "fossils" of ancient transposons/inactivated viruses but are often integrated in active but non-coding transcriptional regions (barely affecting the stability of the genome). TE are considered to be the main contributors to the diversification of lncRNA in vertebrates (Ganesh and Svoboda, 2016). In addition to local repeats, they provide lncRNA with TSS, signals for splicing, polyadenylation, RNA maturation and lncRNA binding sites as well as sub-cellular localization signals and signals for secondary structures specific for binding to certain protein domains; in short, they bring their diversity to the genome, thus participating in the fluidity and rapid evolution of lncRNA (Kapusta *et al.*, 2013; Necsulea *et al.*, 2014).

Finally, permissive transcription of the genome constantly generates RNA molecules that are for the most part rapidly degraded. If these are selected through evolutionary pressure, they can shift from the unstable to stable state and thus give rise to a new lncRNA with new functions.

However, the *de novo* origin of lncRNA remains difficult to determine and is only represented through a handful of examples, such as the Poldi lncRNA, specific to testis and whose origin is a coding but un-translated gene (Heinen *et al.*, 2009).

In humans, the testis and cerebral cortex are the tissues with the richest expression of dedicated non-coding genes. This particularity could contribute to certain phenotypic traits that are unique to the human species, such as a more developed cognitive capacity (Durruthy-Durruthy *et al.*, 2016; Wu *et al.*, 2011).

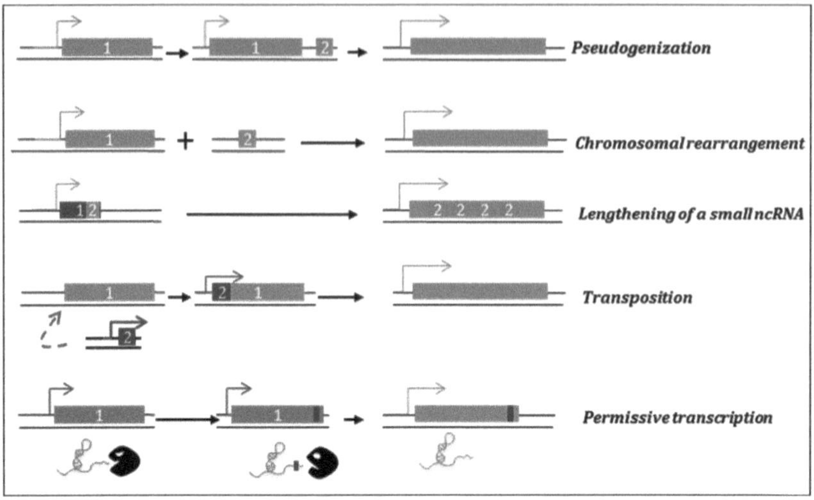

Figure 3.1. *Models of the possible origins of lncRNA. For a color version of this figure, see www.iste.co.uk/morillon/RNA.zip*

NOTES ON FIGURE 3.1.– Either from the pseudogenization of a coding gene; chromosome rearrangements juxtaposing several sequences; the lengthening of a short transcript; the insertion of a sequence through transposition to bring its promoter closer; or an unstable RNA acquiring greater stability following a mutation.

3.1.2. *Conservation and evolution of lncRNA*

The multiplication of genomic and transcriptomic data for several eukaryote models has enabled the analysis of sequence conservation between coding and non-coding genes. These studies have revealed that the human genome is very dynamic and only 8.2% of its DNA sequence is subject to evolutionary constraints (Rands *et al.*, 2014). A remarkable fact is that non-coding genes are amongst the least conserved, with more than 80% of lncRNA families originating from primates (Necsulea *et al.*, 2014).

This discovery raised much skepticism on the functionality and biological pertinence of lncRNA, and resulted in the search for other conservation constraints (Ponting *et al.*, 2009). Although primary sequence conservation criteria may be too restricted in terms of lncRNA, other characteristics such as its structure, function and syntenic loci-based expression constitute multidimensional factors that are more applicable to dynamic studies on lncRNA. Recently, a study on the non-coding transcriptome of 17 different species (16 vertebrates and a sea urchin) showed that although the bulk of non-coding genes generally were not conserved, short conserved sequences could be found at their 5' extremity. This confirmed a higher conservation of TSS and an important level of synteny, in several animal species, between very specific gene networks, notably in development (Hezroni *et al.*, 2015). Effectively, the most conserved are development lncRNA from the lincRNA subgroup.

The conservation of these lncRNA is especially notable in terms of its temporal expression dynamic and gene localization (synteny), suggesting that they are selectively maintained and crucial to the development process (Necsulea *et al.*, 2014; Ulitsky *et al.*, 2011; Washietl *et al.*, 2014).

3.2. Functions of lncRNA in biological diversity

The ever-increasing number of lncRNA identified over the last decade has now largely surpassed that of protein-coding transcripts. The diversity of the non-coding transcriptome is today considered as a strong argument in explaining the phenotypical differences seen across species, given the relatively similar number of coding genes between the same species – 13,985 in fruit flies, 21,009 in the nematode and 23,341 in man (Willingham and Gingeras, 2006).

In 2001, John Mattick and Michael Gagen suggested, for the first time, that non-coding transcripts, as well as introns, constitute an endogenous network enabling dynamic communication from gene to gene. Contrary to basic proteomic circuits, this superior regulation system relies on RNA and operates through intermediary RNA-DNA, RNA-RNA and RNA-protein interactions to encourage the evolution of sophisticated multicellular organisms and the rapid expansion of phenotypic complexity (Figure 3.2). A direct correlation between the proportion of non-coding sequences in the genome and the complexity of the organism has been suggested (Mattick, 2001; Mattick and Gagen, 2001). It is interesting to note that comparative genomics have permitted the identification of several regions of the human genome that are highly divergent compared with other species (Bird *et al.*, 2007 ; Pollard *et al.*, 2006a). These human accelerated regions (HAR) contain many lncRNA and are involved in the acquisition of particular human characteristics over the course of evolution.

In 2006, an initial lncRNA of these regions was shown to be expressed during cortical brain development (Pollard *et al.*, 2006b). Since then, many mutations involved in disease have been identified in these non-coding regions and have been shown to be associated with brain activity regulator elements (Bae *et al.*, 2014). A more recent study showed that mutations of HAR enhancer elements could be involved in the development of autism, thus supporting the hypothesis that certain HAR control human-specific behavioral traits, such as social interaction (Doan *et al.*, 2016). However, although the functionality of non-coding transcripts has been and still remains highly debated, their exhaustive molecular characterization has become a major challenge.

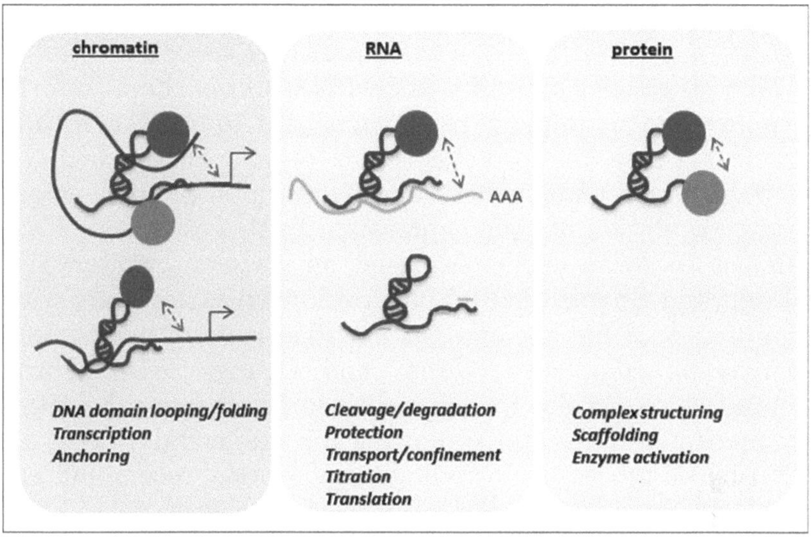

Figure 3.2. *lncRNA regulation mechanisms. For a color version of this figure, see www.iste.co.uk/morillon/RNA.zip*

NOTES ON FIGURE 3.2.– lncRNA can serve as structures/scaffolding for histone modification complexes and induce transcription activation and/or suppression. They can interact directly or indirectly with mRNA and regulate their splicing, degradation or translation. Finally, they can act as miRNA sponges and prevent their action on target mRNA.

3.3. The classified functions of lncRNA

To highlight their regulatory role, lncRNA are often classed based on their function. Several classic activities are now used for their classification: protein complex scaffolding, guides, lures or ribo-suppressors, ribo-activators and sponges, precursors of small ncRNA. Examples of functional classifications that group several lncRNA into sub-classes with a common functional mode will be presented here.

3.2.1. *"Scaffold" lncRNA*

The structural plasticity of lncRNA allows them to form complex and dynamic three-dimensional complexes with a strong affinity for proteins (Guo *et al.*, 2016). lncRNA scaffolds are often epigenetic and transcription control actors in the regulation of gene expression. In this case, a lncRNA can act in *trans* or in *cis* based on its transcription site (Quinn and Chang, 2016). They are known to associate with a multitude of histone or DNA modification and nucleosome remodeling complexes (Quinn and Chang, 2016). These complexes reorganize the epigenetic landscape and the set of chromatin domains, thus allowing for the modulation of all DNA-based processes, including transcription, recombination and RNA repair and maturation (Gonzalez *et al.*, 2015 ; Lee *et al.*, 2016; Yoon *et al.*, 2013; Zheng *et al.*, 2015).

HOTAIR is an example of a scaffold lncRNA that recognizes many targets. HOTAIR forms a secondary structure with four modules that interact, in the nucleus, with the PRC2 and Lsd1/REST/coREST complexes across the 5' and 3' modules, respectively (Tsai *et al.*, 2010), and therefore targets them for genomic positions where they can carry out histone modification and/or transcriptional suppression (Figure 3.3). In the cytoplasm, HOTAIR associates with ubiquitine ligases E3, Dzip3 and Mex3b, facilitating the ubiquitination and proteolysis of their respective substrates, Ataxin-1 and Snurportin-1, in senescent cells (Yoon *et al.*, 2013). Through this example, the high versatility of these molecules, whose function depends largely on factors that interact with them in distinct cell compartments, is understood.

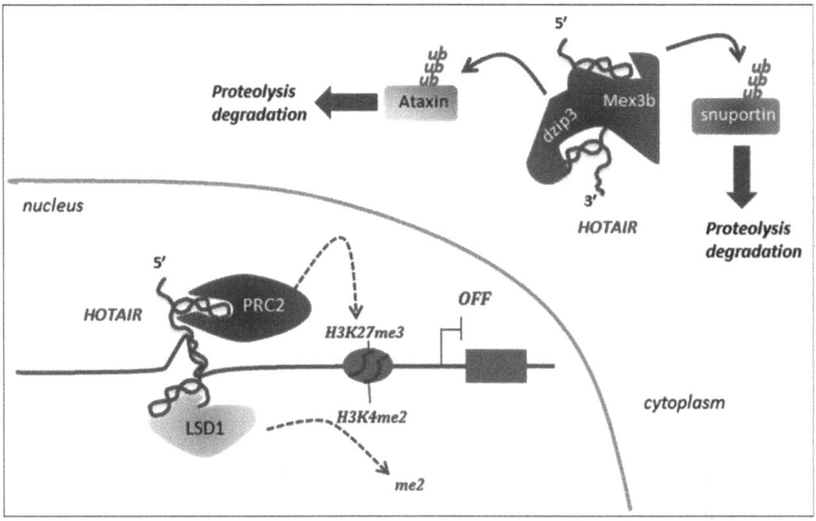

Figure 3.3. *HOTAIR modules and the multiple functions of lncRNA. For a color version of this figure, see www.iste.co.uk/morillon/RNA.zip*

NOTES ON FIGURE 3.3.– HOTAIR includes 2 modules for binding to chromatin PRC2 and LSD1 coREST/REST modification complexes, respectively involved in the methylation of the H3 histone to lysine 27 and in the demethylation of the H3 histone to lysine 4. These two nuclear processes lead to a transcriptional inactivation of genes targeted by HOTAIR. In the cytoplasm, HOTAIR binds to ubiquitine ligases dzip3 and mex3b to accelerate the degradation of Ataxin and snuportin.

3.2.2. *Architect lncRNA (arcRNA)*

arcRNA are a sub-class of scaffold lncRNA, essential in the assembly of specific nuclear sub-structures (Chujo *et al.*, 2016). Currently, five lncRNA are classed as arcRNA. Of these, the NEAT1 lncRNA assembles more than 60 different proteins of which some are RNA-binding proteins while others are transcription factors, all this within a compartment/foyer called paraspeckle (Yamazaki and

Hirose, 2015). arcRNA are heavily enriched in repeated sequences indicating a complex RNA structure, no doubt crucial for their scaffolding function. They can be temporarily regulated by stress, through the course of development or in specific pathologies. arcRNA often sequester regulatory proteins, thus modifying the expression of genes.

3.2.3. *"Guide" lncRNA*

Guide lncRNA are able to recruit RNP complexes in specific chromatin regions. Remarkably, the function of the same lncRNA depends on the biological context (type of cell/tissue, developmental stage, pathology) and often cannot be explained by simple RNA/DNA sequence complementarity. For certain guide lncRNA, the formation of a triple helix structure between DNA and RNA has been proven experimentally, such as the case of Khps1 which binds the CBP/p300 complex to the SPHK1 proto-oncogene (Postepska-Igielska *et al.*, 2015). Another example is the MEG3 lncRNA that guides PRC2's EZH2 sub-unit towards genes regulated by TGFß (Mondal *et al.*, 2015) during epithelial-mesenchymal transition.

3.2.4. *"Decoy" lncRNA*

Decoy lncRNA play the role of ribo-suppression in enzymatic activities through allosteric modification, inhibition of catalytic activity or by blocking binding sites. A classic example of a ribo-suppressor lncRNA is GAS5, which acts as a decoy for a glucocorticoid receptor (GR) by imitating the glucocorticoid response element (GRE) present on genomic DNA. The interaction of GAS5 with the GR prevents it from binding to the genomic GRE template and limits the transcriptional induction of genes regulated by GR, thus controlling metabolism, cell survival and response to apoptotic stimuli (see Figure 5.1) (Kino *et al.*, 2010).

lncRNA can also act as essential ribo-activators or favoring protein activities. One such example is the lnc-DC lncRNA which pushes the phosphorylation and activation of the STAT3 transcription factor (Wang *et al.*, 2014). Others can interact with and regulate the kinase activity of Mediator, thus facilitating the formation of a chromatin loop between the gene promoter and its enhancer, encouraging the activation of transcription (Lai *et al.*, 2013). In addition to the RNA interacting with the mediator, other lncRNA are able to up-regulate transcription. Amongst these is SRA, a steroid receptor activator that interacts with the CTCF and NeST protein, stimulating the activity of a sub-unit of the H3 Lysine 4 methyltransferase histone, a transcription activator (Gomez *et al.*, 2013; Yao *et al.*, 2010).

3.2.5. *Competitive endogenous lncRNA (ceRNA)*

ceRNA, also known as sponge lncRNA, are represented by lncRNA and circRNA that share partial sequence similarity with PCG transcripts. They compete to bind miRNA and for post-transcriptional control (Szczesniak and Makalowska, 2016). lncRNA derived from pseudogenes represent an important source of ceRNA as they are particularly enriched in miRNA response elements, such as the previously mentioned PTENP1 (An *et al.*, 2017). The sub-cellular balance between a ceRNA, one or more miRNA and target mRNA forms a complex network allowing for the precise adjustment of gene expression regulation during adaptation, response to stress and development (Tay *et al.*, 2014).

3.2.6. *miRNA precursor lncRNA*

Many lncRNA include small genes and act as precursors for shorter regulator RNA, especially those involved in the RNAi pathway (mi/si/piRNA). Many lncRNA were functionally identified and studied before their precursor function was known. Such was the case for H19, one of the first lncRNA discovered, and which contains two conserved microRNA, miR-675 and miR-675-5p. In undifferentiated cells, H19 acts as a ribo-activator interacting with and encouraging the activity of the KSRP (K-type splicing regulator protein) RNA-binding protein to prevent myogenic differentiation (Giovarelli *et al.*, 2014). Throughout development and particularly during the differentiation of skeletal muscle, H19 matures into miRNA to ensure post-transcriptional control of Smad anti-differentiation transcription factors (Dey *et al.*, 2014). Certain piRNA clusters have been mapped, as was done for lncRNA, notably in exon regions but also in mobile element-rich regions, thus constituting potential piRNA precursors (Ha *et al.*, 2014). Putative endo-siRNA match with expected inverse hairpin RNA protrusions, but could also originate from any such double stranded RNA-RNA precursor produced through convergent sense-antisense transcription (Carlile *et al.*, 2009; Werner, 2013). Endo-siRNA have been documented in many eukaryotes, including in flies, nematodes and mice.

Besides the potential to form double-stranded structures, the maturation of lncRNA into small RNA molecules could also depend on different mechanisms and cell enzymes such as RNaseP and RNaseZ, as seen in the maturation of MALAT1 into small RNA (Wilusz *et al.*, 2008), during which the possible coexistence of two modes of operation combining a long lncRNA precursor and a small derived RNA adds another layer of complexity to the lncRNA-regulated circuits.

3.3. Classification based on association with biological processes

Non-coding transcriptome studies in the different biological contexts of normal and pathological development have led to the discovery of lncRNA specifically associated with biological states or with specific pathologies. lncRNA expressed differently during replicative senescence represent senescence-associated lncRNA, or SAL (Abdelmohsen *et al.*, 2013). For example, SALNR is able to delay oncogene-induced senescence through its interaction with NF90 (Wu *et al.*, 2015). Hypoxia, a classic characteristic of the tumor micro-environment, induces the expression of many lncRNA, in particular those of the UCR, called HINCUT (Choudhry *et al.*, 2016; Ferdin *et al.*, 2013). Oxidative stress induces the production of stress-induced lncRNA, the si-lncRNA, which accumulate at the polysomic level unlike mRNA whose levels decrease (Giannakakis *et al.*, 2015). Finally, with the progress of clinical and diagnostic studies, an increasing number of lncRNA associated with disease have been characterized. For example, prostate cancer associated transcripts (PCAT) such as PCAT1 have a role in tumor progression (see Figure 5.1) but can also be used as prognostic markers (Prensner *et al.*, 2011).

4

Non-coding RNA in Development

Among the first examples of the best characterized lncRNA that have specific roles in development and that demonstrate robust loss of function phenotypes *in vivo* are those involved in dosage compensation and genomic imprinting (see Figures 1.5 and 1.6). These two processes are necessary for normal development and rely on the establishment of "silent" chromatin to generate the mono-allelic expression of genes specific to mammals (Lee and Bartolomei, 2013).

4.1. Inactivation of the X chromosome

4.1.1. *Identification and expression of XIST*

The identification of XIST as a regulator of X chromosome inactivation in mammals provided one of the first examples of lncRNA directly involved in the formation of suppressor chromatin (Lee and Bartolomei, 2013). The deletion of XIST in mice leads to the loss of X chromosome inactivation and is specifically lethal in female mice (Penny *et al.*, 1996). Various studies carried out in mice and mouse embryonic stem cells demonstrated that, in female cells, XIST acts in *cis* by inducing the formation of inactive heterochromatin for transcription on the X chromosome. Although XIST is required for initiation, it is not needed to maintain X inactivation and its spatio-temporal expression is finely controlled. XIST induces the formation of suppressor heterochromatin, at least in part, by leading PRC2 (see Figure 1.6) to the inactive X (Zhao *et al.*, 2008). However, parallel X inactivation pathways, independent of PRC2, have recently been demonstrated (Nozawa *et al.*, 2013; Schoeftner *et al.*, 2006).

XIST is regulated transcriptionally through a network of pluripotency factors and can also play an important role in differentiation. Effectively, the conditioned homozygote and heterozygote deletion of XIST in mouse hematopoietic stem cells produces an abnormal maturation of hematopoietic progenitors in females (Yildirim *et al.*, 2013), which leads to the development of blood cell cancers and premature death. An abnormal XIST expression has been seen in human cancers, suggesting that the alteration of the X inactivation process could contribute to tumorigenesis.

4.1.2. *The mechanism of XIST regulation*

The interaction of XIST and chromatin can involve, amongst others, the YY1 transcription suppressor that serves as a recruitment platform by binding to the first XIST exon (Jeon and Lee, 2011). What's more, XIST itself is able to recognize the three-dimensional structure of the X chromosome (Engreitz *et al.*, 2013). The expression of XIST is itself controlled by other lncRNA (Lee and Bartolomei, 2013). Amongst these is the Tsix non-coding antisense transcript (see Figure 1.6), which suppresses the expression of XIST by inducing epigenetic repressive modifications on its promoter (Lee and Bartolomei, 2013). The loss of Tsix function *in vivo* results in the ectopic expression of XIST, an abnormal inactivation of X and premature embryonic lethality (Lee, 2000 ; Sado *et al.*, 2001). Additionally, activation of XIST also requires the Jpx62 lncRNA which induces the transcription of XIST by sequestering the CTCF38 repressive transcription factor.

4.2. Genomic imprinting

Imprinted genes generally gather in foci and are marked, epigenetically, based on gender during male and female gametogenesis. They are then switched off by a single parental chromosome in the embryo. The imprinted regions encode different categories of ncRNA, including lncRNA that, in many cases, bind to imprinted regions and are directly involved in transcriptional silencing (Lee and Bartolomei, 2013). These lncRNA are generally long (more than 100 kb) and work in *cis*. The two examples that are best characterized at both the genetic and molecular level are the Kcnqlot1 lncRNA that overlaps with the Kcnq1 gene, and Airn, the antisense of IgF2r. These lncRNA are expressed paternally and function by suppressing coding genes that are juxtaposed in *cis* and involved in the early development of mice (Lee and Bartolomei, 2013). The loss of function of these lncRNA in the embryo is not lethal but presents a growth default in the embryos that inherit the paternal loss of function (Fitzpatrick *et al.*, 2002; Mancini-Dinardo *et al.*, 2006; Sleutels *et al.*, 2002). It is interesting to note that, during the establishment of transcriptional silencing in genes, transcriptional interference can be more efficient than the production of a regulator lncRNA. For example, in embryonic tissues, Airn suppresses the paternal Igf2r locus through a mechanism that does not require an RNA product, but that is based on the continued transcription of Airn, thus interfering (see Figure 2.3) with the recruitment of RNA polymerase II (Latos *et al.*, 2012). On the other hand, in the placenta, Airn RNA recruits EHMT2 to induce the formation of a repressive chromatin (Sleutels *et al.*, 2002). These studies show that a single lncRNA can operate through different mechanisms based on cell type. It is therefore not surprising that the use of *cis*-acting lncRNA is an evolutionarily conserved mechanism and is not limited to complex and multicellular organisms, as is the case for the cryptic unstable transcripts in yeast (Colin *et al.*, 2011).

4.3. Regulation of HOX genes

HOX genes code for a family of conserved transcription factors that regulate the spatial arrangement of the embryonic body and that contribute to the cell differentiation process (Barber and Rastegar, 2010). In mammals, there are 39 HOX genes that are classed into four groups (HOXA, HOXB, HOXC and HOXD), enabling the precise spatio-temporal coordination of their respective expressions. In addition to coding genes, these networks produce certain lncRNA that have similar spatio-temporal expression windows to neighboring coding genes (Rinn *et al.*, 2007). Some of these lncRNA have proven to be directly involved in the regulation of HOX genes.

4.3.1. *lncRNA that act in* cis

HOTTIP is an antisense lncRNA of the 5' end of the human HOXA locus, upstream of HOXA13 and identified in human primary fibroblasts (Figure 4.1). The negative regulation of HOTTIP levels in primary fibroblasts induces the transcription of several genes downstream of HOXA. HOTTIP is conserved in vertebrates and its inactivation in chicken embryos alters the morphology of limbs (Wang *et al.*, 2011b). HOTTIP regulates the expression of HOXA by interacting with the MLL1 histone modification complex and by activating the formation of chromatin loops that connect HOXA gene promoters (Wang *et al.*, 2011b) (Figure 4.1).

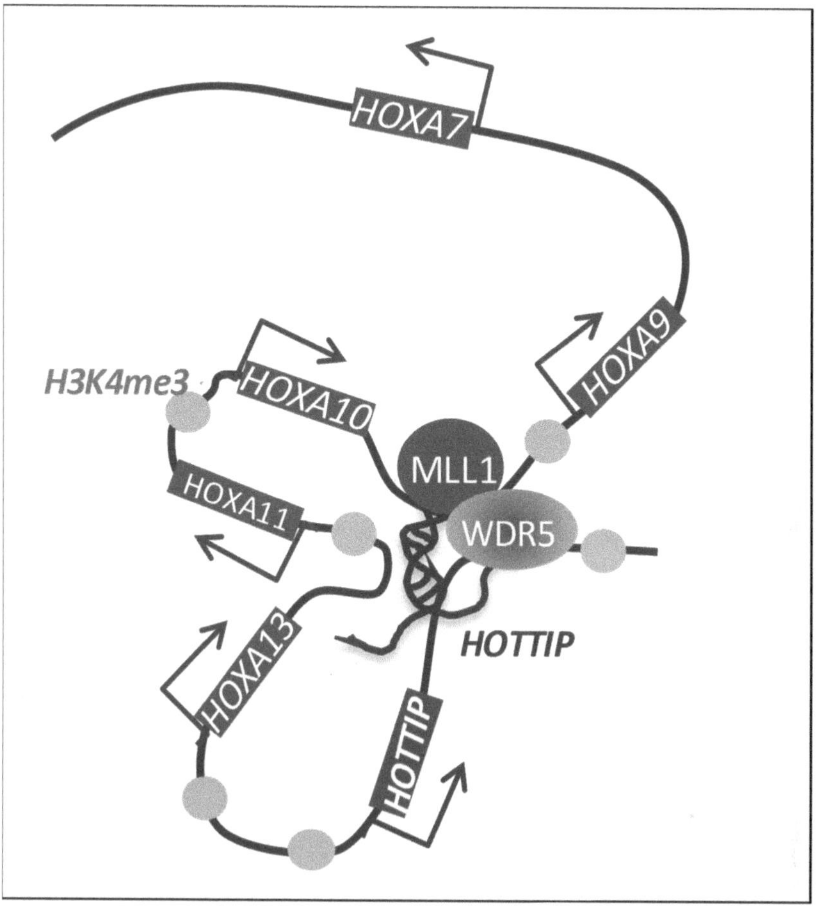

Figure 4.1. Expression regulation through *HOTTIP. The HOTTIP lncRNA recruits the MLL1/WDR5 complex and configures the chromatin to mark the domain with the Histone H3 lysine 4 trimethyl (H3K4me3) mark, and thus allows the expression of adjacent HOX genes. For a color version of this figure, see www.iste.co.uk/morillon/RNA.zip*

An identical mode of action has been described for the mistral (Mira) lncRNA, specific to mice and transcribed from the HOXA locus (Bertani *et al.*, 2011). It positively controls the transcription of two adjacent genes, HOXA6 and HOXA7. Inactivation of Mira in mouse ESCs inhibits the activation of differentiation genes in the germinal layer, suggesting a role in the initiation of cell differentiation (Bertani *et al.*, 2011) but independently from HOXA6 and 7, neither of which are involved in this stage of development.

HOTAIRM1 is a lncRNA identified as the antisense at the 3' extremity of the HOXA locus, specifically expressed in myeloid lineages (Zhang *et al.*, 2009). Inactivation of HOTAIRM1 in myeloid leukemia lines causes the shutdown of the expression of the HOXA 3' downstream coding genes through a mechanism that has not yet been characterized, but that suggests a positive role in gene expression.

4.3.2. *lncRNA that act in* trans

HOTAIR is one of the first described lncRNA to work in *trans* (Rinn *et al.*, 2007). It is antisense to the HOXC genes but acts as a suppressor of the HOXD network, located on a different chromosome. HOTAIR interacts with the PRC2 and LSD1/coREST/REST complex (Figure 3.3), modulating histones and recruiting them on specific target genes. The inactivation of HOTAIR in human fibroblasts leads to an increase in HOXD gene expression. The deletion of the HOTAIR gene does not affect viability in mice but leads to development defects and transformations in the skeletal system (Maamar *et al.*, 2013).

It is interesting to note that the deletion of the whole HoxC locus in mice is lethal but does not lead to skeletal transformations, whereas deletions of its individual components, including HOTAIR, are viable but lead to developmental defects (Schorderet and Duboule, 2011; Suemori and Noguchi, 2000). This observation could reflect the presence of compensation mechanisms amongst the members of the HOXD group or of genes whose functions are antagonist to HOTAIR (Li *et al.*, 2013a), thus underlining the importance of using appropriate models to define the function of lncRNA. Finally, it was discovered that HOTAIR is positively regulated in different cancers as described in more detail in Chapter 5 (Gupta *et al.*, 2010).

4.4. Pluripotency by preventing the initiation of cell differentiation

Several lncRNA are associated with pluripotency and have been either identified as induced during the reprogramming of fibroblasts into induced pluripotent stem cells (iPSCs) (Loewer *et al.*, 2010) or are expressed in mouse (Dinger *et al.*, 2008; Guttman *et al.*, 2009 ; Guttman *et al.*, 2011) or human (Ng *et al.*, 2012) embryonic stem cells. Notably, the expression of these lncRNA correlates with that of the OCT4 (or POU5F1), NANOG and SOX2 genes, essential components of the pluripotency transcription network (Dinger *et al.*, 2008). Additionally, the promoters of these lncRNA are recognized by at least one of these transcription factors (Guttman *et al.*, 2011). Loss of function experiments induces either a release from the pluripotent state or an activation of differentiation programs (Guttman *et al.*, 2011 ; Sheik Mohamed *et al.*, 2010).

Similarly to the modular scaffolding hypothesis described in Chapter 2, the lncRNA-ES1 and lncRNA-ES2 interact with both SOX2 and the SUZ12 Polycomb protein suggesting a recruitment platform model for SUZ12 that switches off the SOX2 neuron targets in pluripotent ESC (Ng *et al.*, 2012) (Figure 4.2).

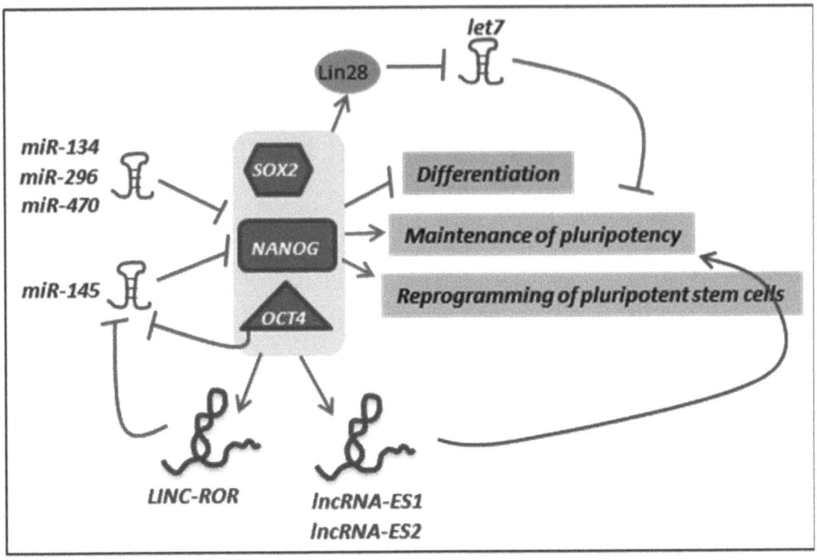

Figure 4.2. Contro*l of pluripotency. Schematic representation of the regulation circuits that involve the pluripotency transcription factors NANOG, SOX2 and OCT4 (control of pluripotency transcription factors: CPTF), the different microRNA (miR-134/296/470/145 and let7) involved, of which mir145 controlled by LINCROR and OCT4 and lncRNAs (LncRNA-ES1/ES2 and LINC-ROR). For a color version of this figure, see www.iste.co.uk/ morillon/RNA.zip*

Among the lncRNA associated with pluripotency, LINC-ROR is constantly enriched with human iPSC, no matter the cell of origin (Loewer *et al.*, 2010). It acts as a ceRNA (see Chapter 3) to regulate the expression of pluripotency transcription factors competing for a linkage to miR-145 (Wang *et al.*, 2013) (Figure 4.2). This clear example perfectly illustrates the way in which regulation pathways between transcription factors, miRNA and lncRNA, are interconnected to control cell fate.

4.5. Brain and central nervous system (CNS) development

The central nervous system (CNS), which contains a wide range of neuronal sub-types and glial cells, is by far the most complex and the most diverse organ in terms of lncRNA (Qureshi and Mehler, 2012) and it seems that their number is linked with evolutionary complexity.

4.5.1. *The abundance of lncRNA in the nervous system*

MALAT1, originally identified in pulmonary adenocarcinoma, is among the most abundant lncRNA in the nervous system. It is described as heavily expressed in different types of neurons in mice (Bernard *et al.*, 2010). MALAT1 is found in nuclear foci and appears to regulate synapse formation (Bernard *et al.*, 2010). Repression of MALAT1 in mouse hippocampus neurons results in a decrease in synapse density and a reduction in dendrite growth (Qureshi and Mehler, 2012). However, the loss of MALAT1 function is not critical to prenatal and postnatal mouse development (Zhang *et al.*, 2012), suggesting perhaps a functional redundancy that might have been lost through the cell lineage. Overall, the exact mechanisms of MALAT1 activity, within the processes of tumor progression and neuronal differentiation, remain to be determined.

4.5.2. *lncRNA associated with the expression of neuronal development proteins*

Analysis of the brain transcriptome has shown that many lncRNA are either specific to primates, or to humans (Xu *et al.*, 2010), suggesting that lncRNA could be determining factors in the evolution of the human brain (Lindblad-Toh *et al.*, 2011; Pollard *et al.*, 2006b). The HAR1A lncRNA that has evolved the fastest since the divergence of man from great apes (see Chapter 2), has a level of expression that correlates with that of reelin, a crucial protein in brain development, suggesting an interaction in the gene pathway that remains to be elucidated.

Other classes of lncRNA expressed in the brain seem highly conserved, from birds to mammals, and have similar spatio-temporal expression profiles that indicate primitive and conserved roles in brain development (Dorus *et al.*, 2004). These lncRNA generated from ultraconserved regions (UCR) have shown themselves to be transcribed from genetic loci where they often overlap, or are antisense to, genes coding for key development regulator proteins (Feng *et al.*, 2006). Such lncRNA modulate the activity of their neighboring genes by acting as molecular scaffolds for the recruitment of specific factors (Bond *et al.*, 2009; Feng *et al.*, 2006). For example, Dlx6os1, antisense of Dlx6 and downstream of Dlx5, codes for transcription factors and is expressed in the brain during anterior ventral development. It may play a role in craniofacial development. Dlx6os1 controls the expression of Dlx5, Dlx6 and Gad1 (which codes for an enzyme responsible for the synthesis of γ-aminobutyric acid: GABA) – through *cis* and *trans* mechanisms of action (Feng *et al.*, 2006). The loss of function of Dlx6os1 in mice produces a neuron-specific phenotype with a reduced number of "GABAgenic" interneurons in the early hippocampus, which indicates the crucial role of Dlx6os1 in neuronal activity *in vivo* (Bond *et al.*, 2009; Feng *et al.*, 2006).

4.5.3. *lncRNA of the retina*

It is interesting to note that several lncRNA have revealed themselves to be specifically expressed in the retina, a highly specialized part of the CNS. The retina is a valuable tissue type for *in vivo* studies since the loss of function of genes can be obtained through the local administration of RNA interference reagents (RNAi), unlike the germline genetic modifications that are necessary for many *in vivo* studies of the CNS.

4.5.4. *The circRNA of the brain*

CDR1AS (also known as ciRS-7) has been identified in the human brain (see Figure 2.3), as well as in mouse pyramid neurons and interneurons (Hansen *et al.*, 2011). It is generated from the non-canonical splicing of an antisense transcript of the CDR1 gene, linked with cerebellar degeneration. This circRNA functions as a sponge for miR-7 and thus the endogenous regulation of its targets (Hansen *et al.*, 2013 ; Memczak *et al.*, 2013). Zebrafish embryos that express ectopic CDR1AS develop cerebral defects similar to the phenotype seen for the loss of function of miR-7 (Memczak *et al.*, 2013), suggesting that circRNA may also have roles in neuronal function and in neurological issues (Hansen *et al.*, 2013 ; Memczak *et al.*, 2013).

In conclusion, the multidimensional functions of lncRNA appear appropriate for the complex regulation requirements of the CNS, and other studies on lncRNA will almost definitely reveal their involvement in even more complex cerebral functions as well as neurodegenerative syndromes.

4.6. Development of other organs

In addition to their extensive roles in brain development, lncRNA are known to for their roles in the development of the various organs and tissue types described below.

4.6.1. *The heart*

The best examples of the importance of lncRNA in the development of organs are the lncRNA braveheart (Bvht or Grn20748) and Fendrr (RNA regulator adjacent to Foxf1), involved in cardiac development in mice (Grote *et al.*, 2013; Klattenhoff *et al.*, 2013). Both these lncRNA are expressed in the mesoderm, from which the heart originates. Inactivation of Bvht through RNA interference in mouse ESc and neonatal cardiomyocyte cultures affects the expression of cardiac genes and alters the development of mature cardiomyocytes, thus suggesting Bvht's potential role in the regeneration of cardiac tissues after injury (Klattenhoff *et al.*, 2013).

The interaction of Bvht with PRC2 suggests an epigenetic regulation mechanism. As for Fendrr, the simple deletion of the gene (and not the reduction of its expression) leads to embryonic lethality caused by cardiac insufficiency and deficits in the body walls (Grote *et al.*, 2013). Fendrr interacts with chromatin repressive complexes (such as PRC2) in mouse embryos. Chromatin immunoprecipitation (ChIP) analyses after the suppression of Fendrr have shown changes in the activity of PRC2 on its target genes, suggesting a mechanism based on chromatin reallocation (Grote *et al.*, 2013). Unlike Bvht, Fendrr has human orthologue transcript, FENDRR, which is also associated with PRC2 (Khalil *et al.*, 2009) in a similar way to murine Fendrr, indicating functional conservation.

4.6.2. *The muscles of the skeleton*

One of the first lncRNA identified in myogenesis is Linc-MD1, expressed during the differentiation of mouse myoblasts (*in vitro*). It controls the progression of the early to the late phase of muscle differentiation by working as a ceRNA. Essentially, by titrating miR-133 and miR-135, it regulates the expression of MEML1 and MEF2C, two transcription factors that activate late muscular differentiation genes (Cesana *et al.*, 2011). LINC-MD1 is conserved from mice to men (Twayana *et al.*, 2013) and its expression is highly reduced in myoblasts of patients with

Duchenne muscular dystrophy (Cesana *et al.*, 2011). It is interesting to note that in cells, the artificial increase in levels of LINC-MD1 enables the re-establishment of muscular differentiation *in vitro* (Cesana *et al.*, 2011) (Figure 4.3), opening new potential pathways for a therapeutic approach.

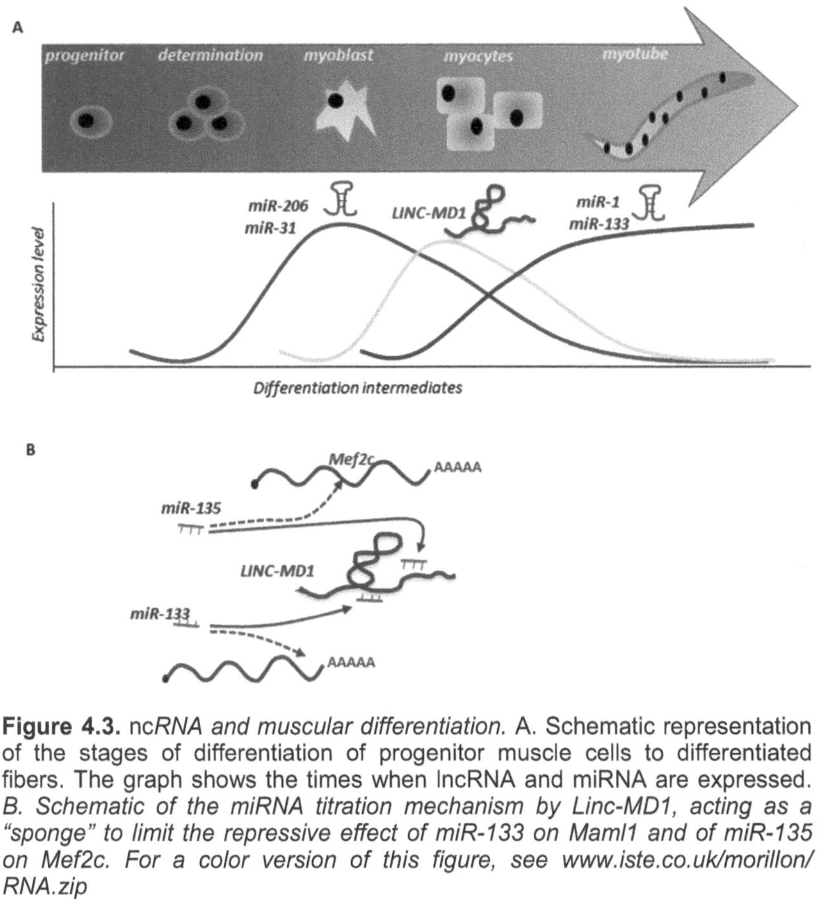

Figure 4.3. nc*RNA and muscular differentiation.* A. Schematic representation of the stages of differentiation of progenitor muscle cells to differentiated fibers. The graph shows the times when lncRNA and miRNA are expressed. *B. Schematic of the miRNA titration mechanism by Linc-MD1, acting as a "sponge" to limit the repressive effect of miR-133 on Maml1 and of miR-135 on Mef2c. For a color version of this figure, see www.iste.co.uk/morillon/ RNA.zip*

More recently, it was shown that the H19 lncRNA involved in parental genome imprinting is not only expressed in embryos during development, but also in adult muscle. A decrease in H19 sets off the premature differentiation of muscles, with a phenotype identical to that seen with the over-expression of the let-7 miRNA. As H19 is known for titrating let-7, it has been suggested that H19 inhibits the activity of let-7 in the same way during premature differentiation (Kallen *et al.*, 2013).

DBE-T is another lncRNA linked to neuromuscular degeneration. It is expressed in patients who suffer from facio-scapulo-humeral muscular dystrophy (FSHD). DBE-T recruits the histone-lysine N-methyltransferase ASH1L (an enzyme component of the MLL1 complex), translating into the abnormal transcriptional activation of the FSHD locus in these patients (Cabianca *et al.*, 2012).

4.7. Development of skin, blood and adipose cells

Other lncRNA have been identified in the epidermis such as TINCR, the most highly activated during the differentiation of keratinocytes (Kretz *et al.*, 2013). The suppression of TINCR leads to an epidermis with no terminal differentiation ultrastructure, such as the absence of keratohyaline granules and intact lamellar bodies.

TINCR also binds to STAU1 and targets mRNA containing a “TINCR box” template of 25 nucleotides, leading to their stabilization and overexpression (Figure 21), such as KRT80 (keratin 80), showing a positive activity on cell differentiation (Kretz *et al.*, 2013). TINCR is also involved in cell proliferation during tumor progression (see Figure 5.1).

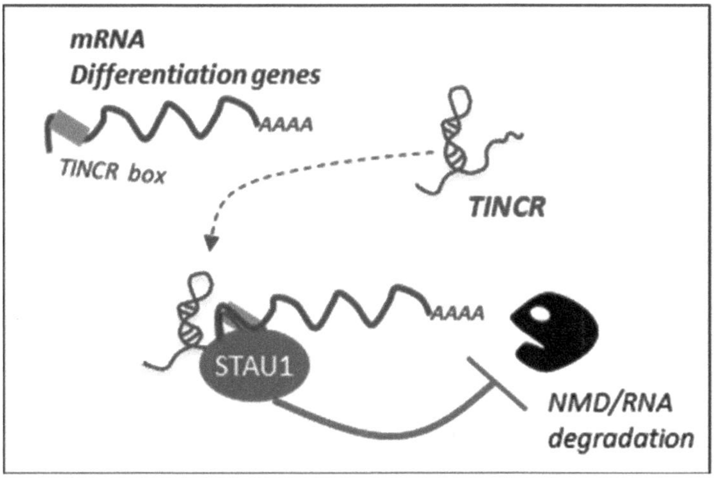

Figure 4.4. *Regulation through TINCR. TINCR binds through sequence complementarity to transcribed mRNAs of cell differentiation genes (containing the "TINCR" box) and inhibits the RNA degradation activity of STAU1 (dependent on the NMD). The stabilized mRNA are therefore expressed. For a color version of this figure, see www.iste.co.uk/morillon/RNA.zip*

lncRNA have also been identified in hematopoiesis and adipogenesis (Hu *et al.*, 2011 ; Sun *et al.*, 2013) as lincRNA-EPS. Decreases in their expression in erythroid progenitors in mice blocks differentiation and encourages apoptosis by inhibiting the expression of the PYD pro-apoptotic genes, through a still poorly understood mechanism (Hu *et al.*, 2011).

5

Long Non-coding RNA and Cancer

Cancer is fundamentally a genetic illness that modifies cell information to alter cell homeostasis and promote growth. The discovery of the universal genetic code for the expression of proteins based on genes allowed for a better understanding of how such mutations lead to cancer. In addition, the sequencing of gene exon regions (exome) allowed the identification of many mutations coding for proteins susceptible of being treated in a targeted manner. However, the coding genome represents less than 2% of all sequences and it has become apparent that abnormalities in the non-coding genome result in phenotypes that are equally important for cancers. As detailed in the previous chapters, one of the most unexpected conclusions from the genomic era of biology is the transcription of non-coding regions of the genome (Morris and Mattick, 2014).

The long list of non-coding RNA runs into the tens of thousands, of which many are uniquely expressed in differentiated tissues or in specific types of cancer (Iyer *et al.*, 2015). However, the function of the large majority of these transcripts remains to be identified. Although the discovery of the universal genetic code provided researchers with the tools to recognize the manner in which genetic mutations lead to functional defects in proteins, RNA biology has, up until now, failed to help understand how the sequence affects function.

lncRNA associated with genomic alterations linked to a cancer

lncARN	Type of cancer	Genomic alteration
PVT1	Colon	8q24 amplification
PCAT-1	Prostate	8q24 snp
CCAT2	Colon	8q24 snp
PTCSC3	Thyroid	Rs944289
HULC	Hepatic	Rs7763881
ANRIL	Several	9p21.3 snp
TERC	Oral cavity	3q26 amplification
GAS5	Hepatic	Rs145204276
SAMMSON	Melanoma	3p13-3p14

lncRNA used in diagnostics

lncRNA	Type of cancer	Tissues for sampling
H19	Gastric	Blood
HULC	Hepatic	Blood
AA174084	Gastrc	Gastric secretions
PCA3	Prostate	Urine
SeCATs	Leukemia (sezary)	Tumor
SPRY4-IT1	Melanoma	Tumor

lncRNA used in prognostics

lncRNA	Type of cancer	Information
FAL1	Ovarian	Bad prognosis
HOTAIR	Breast	Increased risk of metastasis
HOTTIP	Hepatic	Increased risk of progression
MEG3	Meningioma	Associated with grade
NBAT-1	Neuroblastoma	Good prognosis
NKILA	Breast	Decreased risk of metastasis
SCHLAP1	Prostate	Increased risk of metastasis

lncRNA for personalized medicine

lncRNA	Type of cancer	Therapeutic agent
CCAT1	Colon	Inhibitor BET
HOTAIR	Ovary	Platinum-based chemotherapy

Other incRNA detected in cancerous tissues

lncRNA	Type of cancer
PCNA-AS1	Liver
sox2OT	Lung
IncARN-ERIC	Uterine
CAI2	Brain
GHET1	Stomach
HIF1A-AS2	Breat, kidney
XIST	Ovary, breast, kidney, colon
NCRNA	Colon, bladder, brain
SRA-1/SR1	Breast
PTEP1	Lung, melanoma, uterine, leukemia
TUC338	Liver
asNOS2A	Brain
UCA1	Bladder, tongue, breast, colon
PCGEM1	Prostate
LOC285194	Bone, colon, esophagus
PRNCR1	Prostate, colon
MALAT1	Several

Table 5.1. *Current list of lncRNA involved in different cancers, with genomic alteration identified for the diagnosis, prognostics and those identified in cancerous tissues without clinical study*

However, recurrent molecular methods for lncRNA are in development, and emerging technologies are developing researchers' abilities to functionally analyze lncRNA, in particular those associated with cancer. From this perspective, this chapter is an overview of the current state of identification of new biomarkers in cancer phenotypes, their molecular roles in cell processes and in the physiopathology of this disease (Table 5.1).

From this point of view, several examples of lncRNA can be associated with the hallmarks of cancer progression, such as invasion, immortality, longevity, proliferation, apoptosis and growth suppressors (Figure 5.1).

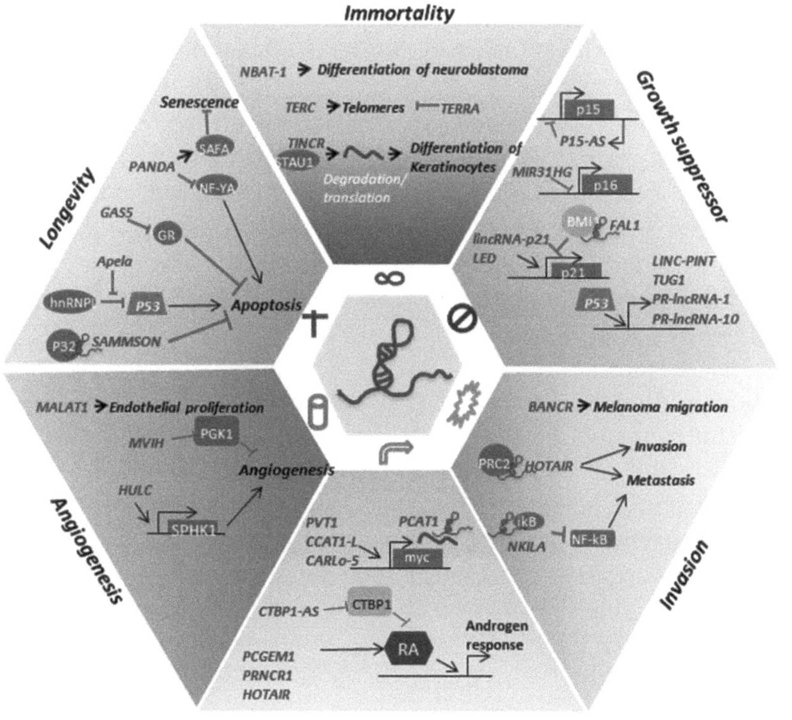

Figure 5.1. *Examples of lncRNA involved in the hallmarks of cancer. The 6 fundamental aspects of carcinogenicity (angiogenesis, proliferation, invasion, growth suppressor, immortality and longevity) are represented with the known examples of lncRNA involved (target or regulator element). For a color version of this figure, see www.iste.co.uk/morillon/RNA.zip*

5.1. Identifying the lncRNA signals in cancer transcriptomes

It is now admitted that mutations in the non-coding genomes are major determinants of human illness (Maurano *et al.*, 2012). Effectively, recurrent somatic non-coding mutations (Li *et al.*, 2015b; Melton *et al.*, 2015), epigenetic alterations (Kundaje *et al.*, 2015), or modifications in the total number of copies (Beroukhim *et al.*, 2010) are all involved in multiple types of cancer. Mutations in regulatory DNA can heavily affect transcription by modifying the activity of the enhancer and the promoter or by modulating the chromatin states, leading to a different expression of mRNA in cancer. Remarkably, several loci that are frequently mutated in cancer bear ultra-conserved non-coding sequences that are deregulated in cancer (Calin *et al.*, 2007). SNP, modifications in the number of copies or mutations in the non-coding genome, can radically alter the transcription of mRNA (Table 5.1). The 8q24 locus is particularly populated with SNP associated with cancer, of which several are involved in the expression of lncRNA associated with cancer, including CCAT2 in colorectal cancers and PCAT-1 in prostate cancer (Ling *et al.*, 2015).

The catalog of non-coding genes has increased considerably over the course of the last few years, in large part due to the identification of whole genome transcription (Djebali *et al.*, 2012; Harrow *et al.*, 2012; Iyer *et al.*, 2015) arising from enhancer (Kim *et al.*, 2011), promoter (Seila *et al.*, 2008) and intergenic sequence activity. These discoveries depend largely on the emergence of new methods of transcriptome assembly and the identification of new RNA (Guttman *et al.*, 2010; Mattick and Rinn, 2015). It is now widely admitted that lncRNA serve as signals of specific cell states or as active readouts of cell program (Wang and Chang, 2011) and may thus allow for the identification of cell pathologies such as cancer, the provision of a prognosis or even information on therapeutic options for patients with cancer (Table 5.1).

The over-expression of the HOTAIR lncRNA in the early stages of breast cancer resected through surgery is highly predictive of metastatic progression and overall survival (Gupta *et al.*, 2010). Subsequent studies have shown that the deregulation of HOTAIR is associated with cancer progression in 26 types of human tumor (Bhan and Mandal, 2015) and can predict the variable sensitivity of patients with ovarian cancer to two platinum-based chemotherapies, which may then be used to guide a clinical decision (Teschendorff *et al.*, 2015).

Differential expression analysis of genes in human prostate cancers identified PCA3 as a specific biomarker of prostate cancer (Bussemakers *et al.*, 1999). PCA3 has since been approved by the Food and Drug Administration (FDA) for the diagnosis of prostate cancer, the first instance of an RNA-based test approved by the FDA. As a biomarker of prostate cancer in urine, it has become a useful and non-invasive test for prostate cancer with an improved specificity, positive predictive value and negative predictive value compared with the prostate-specific antigen PSA test (Wei *et al.*, 2014).

Similarly, analyses of gastric secretions in patients with gastric cancer have permitted the identification of lncRNA-AA174084 as a biomarker able to differentiate gastric cancer and benign gastric epithelium issues (Shao *et al.*, 2014). Recent approaches to the analysis of genomes has enabled the identification of millions of lncRNA that are transcribed differently in normal tissues and in tumor tissues of the same organ (Brunner *et al.*, 2012; Iyer *et al.*, 2015; Yan *et al.*, 2015), suggesting an enormous potential for the further development of biomarkers for specific cancer histology.

The expression profiles of lncRNA can also identify clinically pertinent sub-types of cancer that will predict the behavior of the tumor and the prognosis of the disease (Du *et al.*, 2013).

The over-expression of the SChLAP1 lncRNA in men having undergone radical prostatectomy has been associated with an increased likelihood of development of metastatic progression, similar to the risk associated with high-grade cancer (Prensner *et al.*, 2014). Other studies have identified lncRNA whose expression is specific to the primary tumor stages of prostate cancer or associated with the metastasis of lymphatic ganglions (Bottcher *et al.*, 2015).

Furthermore, recent reports indicate the lncRNA can also predict the responsiveness of patients to cancer therapy types (personalized medicine). The expression of HOTAIR in ovarian cancer is predictive of a mediocre survival following carboplatin treatment (Teschendorff *et al.*, 2015). The validation of these observations could provide new biomarkers to guide the selection/composition of diets containing platinum for patients with ovarian cancer.

5.2. lncRNA, "drivers" of the cancer phenotype

Cancer consists of myriad phenotypes resulting (Hanahan and Weinberg, 2000, 2011) at the same time from a dysfunction in intracellular regulation pathways and from intercellular communication generating a tumor microenvironment.

During cancer development, intracellular signaling networks are modulated to support proliferation, reduce cell apoptosis and differentiation signals, thus improving longevity and promoting motility. During these stages, many lncRNA have been identified as playing a role in these specific cancer phenotypes or as the "hallmarks of cancer" (Figure 5.1).

5.2.1. *LncRNA, new tumor suppressors?*

H19 and XIST were amongst the first lncRNA identified in mammals. As mentioned in the previous chapters, they are exclusively expressed in the maternal allele. H19 is strongly expressed during embryonic development following which its expression level decreases in most tissues. The loss of parental imprinting from the H19 gene locus provokes its over-expression in cancers of the esophagus, colon, liver and spleen. XIST plays an important role in the inactivation of the X chromosome in females. It has been shown that XIST could also have a tumor suppressor role in hematopoietic cancers in mice (Eades *et al.*, 2014; Gibb *et al.*, 2011).

Recently, it has been noted that many other lncRNA play important roles in the modulation of tumor suppression pathways and in cell growth. The pairing of the transcription of certain lncRNA and the dynamics of the cell cycle (Hung *et al.*, 2011) or entry into senescence (Abdelmohsen *et al.*, 2013) positions them as excellent candidates for tumor regulation. Several lncRNA regulate the expression of well characterized tumor suppressors, such as those from the CDKN2A/CDKN2B locus that contain p15INK4b, p16INK4a and p14ARF. The non-coding antisense p15-AS transcript induces the shutdown of the p15INK4b tumor suppressor by promoting the formation of silent heterochromatin (Yu *et al.*, 2008) and the raised expression of p15-AS is associated with a weak expression of p15INK4b in leukemic cells. The MIR31HG lncRNA recruits Polycomb group proteins to the INK4A locus to suppress its transcription during normal growth, but is excluded from the INK4A locus during senescence (Montes *et al.*, 2015). The expression of the TCF21 tumor suppressor is activated by its antisense RNA, which recruits the GADD45a complex on its promoter to facilitate DNA demethylation and transcription (Arab *et al.*, 2014).

The regulation of the p53 tumor suppressor pathway through lncRNA is the subject of particularly intense interest

(Figure 5.2). The MEG3 RNA, expressed from a maternal allele, binds to p53 and activates p53-dependent transcription of a sub-set of genes (Zhou *et al.*, 2007). Furthermore, the number of lncRNA regulated themselves by p53 increases day by day, thus indicating a generalized implication of lncRNA downstream of p53 activation (Huarte *et al.*, 2010; Hung *et al.*, 2011; Marin-Bejar *et al.*, 2013; Sanchez *et al.*, 2014; Younger *et al.*, 2015; Zhang *et al.*, 2014). The genome level profiling of enhancer lncRNA regulated by p53 has enabled the identification of the LED RNA that interacts with and activates super-enhancers, including that of the CDKN1A gene, a major player in stopping the cell cycles through p53. The LED RNA is switched off in certain leukemia cells for which p53 is normal, which indicates a possible role in tumor suppression for this lncRNA (Leveille *et al.*, 2015).

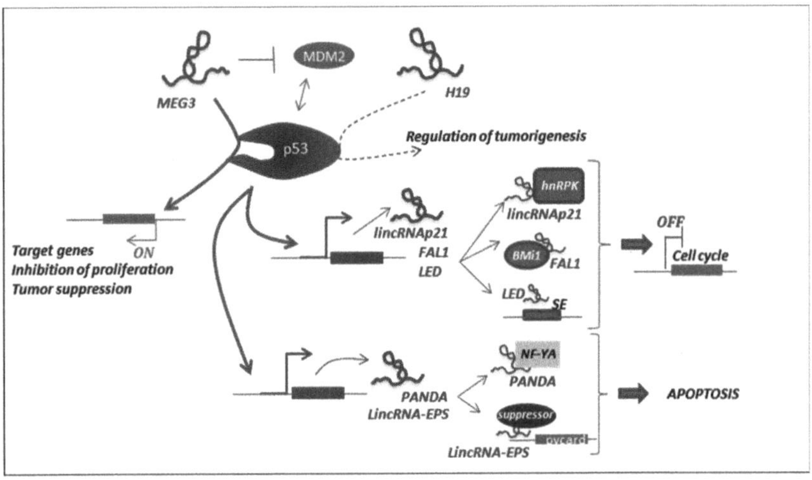

Figure 5.2. *Interrelations between p53 and lncRNA.*
For a color version of this figure, see www.iste.co.uk/morillon/RNA.zip

NOTES ON FIGURE 5.2.– p53 interacts with the Med3 and H19 lncRNA to directly control the transcription of genes involved in tumorigenesis, cell proliferation and tumor suppression. p53 activates the transcription of the lincRNAp21, FAL1 and LED lncRNA that either interact with proteins that control the cell cycle or DNA Super Enhancer (SE) regulator elements. Finally, p53 activates the transcription of PANDA and LincRNA-EPS, forming complexes with NF-YA and a transcription suppressor of the pycard gene involved in apoptosis, respectively.

The lncRNA-p21, induced by DNA damage and in a p53-dependent manner, interacts with hnRNPK to regulate CDKN1A and stops the cell cycle (Dimitrova *et al.*, 2014; Huarte *et al.*, 2010). Similarly, p53 activates the FAL1 lncRNA, located on a chromosome 1 region and frequently amplified in cancers, recruits the BMI-1 suppressor chromatin protein onto multiple genes, including CDKN1A, to promote the proliferation of tumor cells (Hu *et al.*, 2014b).

Other p53 activities are also regulated by the PANDA lncRNA. The latter is able to limit apoptosis induced by DNA damage by binding to the NF-YA transcription factor and blocking its recruitment to pro-apoptotic genes (Hung *et al.*, 2011). PANDA has also been shown to regulate senescence by interacting with SAFA and PRC1 (Puvvula *et al.*, 2014). Finally, lncRNA-EPS suppresses the expression of the pycard gene during erythropoiesis and inhibits the apoptosis of red blood cell progenitors. Effectively, PYCARD is an activator of intracellular caspases for apoptosis (Hu *et al.*, 2011).

5.2.2. *Proliferation circuits*

5.2.2.1. *LncRNA and hormonal pathways*

Many lncRNA, including BCAR4, are downstream targets of chemokines and hormonal pathways (Xing *et al.*, 2014). In T-cell acute lymphoblastic leukemia, the Notch1 oncogene promotes, in part, cell growth by inducing the expression of the LUNAR1 lncRNA to positively regulate the expression and signaling of insulin growth factor 1 receptors (Trimarchi *et al.*, 2014). Androgen signaling in prostate cancer also relies on certain lncRNA, involved in cell proliferation, that work through direct interactions by activating androgen receptor(s) (Yang *et al.*, 2013b; Zhang *et al.*, 2015) or by inhibiting them (Takayama *et al.*, 2013).

5.2.2.2. *LncRNA and the MYC oncogene*

Amplification of the 8q24 locus is a well characterized oncogene event in many types of malignant tumor, which leads to the amplification of MYC. A lot of work now involves the activity of several lncRNA in Myc-dependent cancers (Figure 5.3). For example, in Burkitt's lymphoma, the PVT1 lncRNA is located at the t(2/8) translocation stop point which, when produced, positions the human immunoglobulin amplifier at the PVT1-MYC locus level and activates its expression.

In a mouse model of Myc oncogenesis, amplification into a single copy of Myc is not sufficient to improve the formation of tumors, whereas the amplification of a multi-gene segment encompassing Myc and the Pvt1 lncRNA promotes efficient tumor development (Tseng *et al.*, 2014). In addition, the co-amplification of PVT1 and MYC increases the quantity of Myc protein, as though the lncRNA were also involved in the translational regulation of Myc. As for the latter, the transcription of Myc can be activated in *cis* by the CCAT1 lncRNA (also called CARLo-5), in turn associated with colon cancer, by facilitating the long distance interaction between Myc and an amplifier/enhancer element (Kim *et al.*, 2014; Xiang *et al.*, 2014).

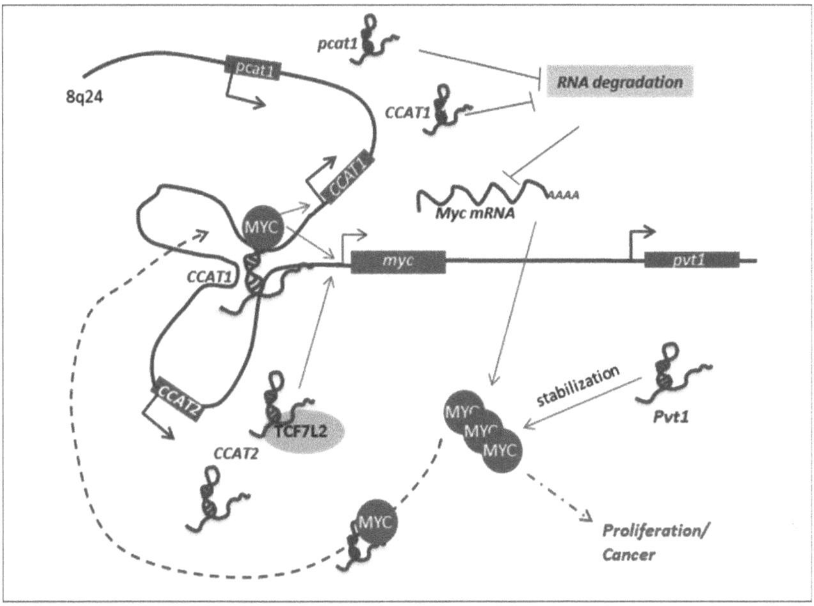

Figure 5.3. *Molecular interactions of the lncRNA-MYC network in the 8q24 genomic region. For a color version of this figure, see www.iste.co.uk/morillon/RNA.zip*

NOTES ON FIGURE 5.3.– The CCAT1 lncRNA activates the expression of MYC both at the transcriptional (through chromatin interactions) and post-transcriptional (inhibition of RNA degradation) levels. The CCAT2 lncRNA stimulates the transcription of MYC by interacting with the TCF7L2 transcription factor. The PCAT1 lncRNA prevents the degradation of the MYC RNA, PVT1 inhibits the degradation of the Myc protein which is important for cancer proliferation. Myc is itself a coactivator of CCAT1 transcription in a feedback loop system. The PCGEM1 lncRNA also appears to interact with Myc in prostate cancer to form transcription activator chromatin structures.

The PCGEM1 lncRNA, specific to the prostate and also located in the 8q24 locus, binds to Myc (Figure 5.3) and improves Myc-dependent transcriptional activation for several genes involved in the metabolic processes necessary for the growth of cancerous cells in the prostate (Hung *et al.*, 2014). Myc in turn regulates the transcription of many lncRNA (Zheng *et al.*, 2014) that can themselves regulate the progress of the cell cycle (Kim *et al.*, 2015).

5.2.3. *Cell longevity circuits*

5.2.3.1. *LncRNA and apoptosis*

The selective advantage of tumor cells is linked to the maintenance of telomeres, to the tolerance of nutritional stress and, for certain cancers, to the preservation of a population of undifferentiated tumor cells. The Gas5 lncRNA is induced in cells arrested following a nutrient deficiency or the withdrawal of growth factors. Gas5 blocks the expression of the gene that responds to glucocorticoids by binding to the DNA binding domain of the glucocorticoid receptor (GR) and by acting as a lure (Hudson *et al.*, 2014; Kino *et al.*, 2010). This GR blockage decreases the expression of the apoptosis cell inhibitor (Kino *et al.*, 2010), whose role is the induction of cell death under stress conditions in normal cells. However, the suppression of Gas5 in cancer cells in humans, compared to adjacent normal mammary tissue, can promote the increased viability of breast cancer cells in a tumor microenvironment lacking in nutritional elements (Mourtada-Maarabouni *et al.*, 2009).

5.2.3.2. *LncRNA and telomeres*

The immortality of tumor cells requires the maintenance of telomeres to prevent the establishment of replicative senescence. The (non-coding) RNA component of telomerase (TERC) is a fundamental, structural and catalytic part of the complex and carries hexanucleotide repeats that form the sequence of telomeres (Feng *et al.*, 1995). The nature of the TERT RNA has been shown as an element predictive of neoplasm progression in the pre-malignant oral cavity (Dorji *et al.*, 2015).

More recently, it has been shown that sub-telomeric and telomeric DNA sequences are transcribed into lncRNA called TERRA, with a role in the maintenance of telomeres (Rippe and Luke, 2015). In terms of mechanism, TERRA regulates the exchange of the single strand DNA binding proteins RPA and POT-1 and, consequently, the recovery of telomeres and their stability (Flynn *et al.*, 2011b).

5.2.3.3. *LncRNA and genome stability*

A recent report also identified a role for the NORAD lncRNA in the maintenance of genome stability (Lee *et al.*, 2016). Since Bovery described aneuploidy in urchins and its role in the promotion of cancers, chromosomal abnormalities have become one of the determining characteristics of cancers (Holland and Cleveland, 2009). NORAD, an abundant lncRNA, sequesters PUMILIO proteins far from their mRNA targets, including those of genes involved in mitosis, DNA repair and replication. PUMILIO proteins negatively regulate these mRNA by decreasing their stability and translation. Thus, mouse cells in which NORAD has been deleted (by mutation) present a hyperactive PUMILIO complex and develop genomic instability and a deleterious aneuploidy.

5.2.3.4. *LncRNA and cancerous stem cells*

lncRNA are widely involved in maintenance and differentiation circuits of stem cells (Flynn and Chang, 2014). Cancer cells can hijack these circuits to modulate gene expression and develop themselves. Thus, LncTCF7 recruits the SW1/SNF complex to the TCF7 promoter and activates Wnt signaling in order to maintain the self-renewal of stem cells in liver cancer (Wang *et al.*, 2015).

Cutaneous cancer cells with squamous cells suppress the TINCR lncRNA. The latter is necessary in the differentiation of keratinocytes, through interaction with STAU-1 and the stabilization of the mRNA associated with differentiation (Figure 4.4) (Kretz *et al.*, 2013).

The NBAT-1 lncRNA has also been observed to promote neuronal differentiation in neuroblastoma cells, by regulating the NRSF/REST neuron-specific transcription factor. Suppression of this lncRNA is therefore associated with a high risk of neuroblastoma (Pandey *et al.*, 2014).

5.2.4. *Cell motility circuits*

The over-expression of MALAT1, strongly and remarkably conserved through evolution, has revealed a high risk of metastatic progression in patients with early stage lung cancer (Ji *et al.*, 2003). Loss of function of MALAT1 in mice showed that this is a non-essential gene in development and in the normal homeostasis of adult tissues (Nakagawa *et al.*, 2012; Zhang *et al.*, 2012). Switching off MALAT1 in mouse pulmonary carcinoma cells alters the *in vitro* cell mobility and slows its metastatic progress through the animal (Gutschner *et al.*, 2013), suggesting that the over-expression of MALAT1 in cancer could lead to gain-of-function phenotypes not observed during normal development of tissues of homeostasis.

Many lncRNA associated with cancer have been implicated in the regulation of invasion and the progression of metastases (Flockhart *et al.*, 2012; Hu *et al.*, 2014c; Huarte, 2015). TGFB, a growth factor, induces the expression of lncRNA-ATB in hepatocellular carcinoma (HCC) cells which facilitates epithelial-mesenchymal transition (EMT), cell invasion and organ colonization by HCC cells through distinct RNA-RNA interactions (Yuan *et al.*, 2014). Effectively, the lncRNA-ATB binds in a competitive manner to miR-200 to activate the expression of ZEB1 and ZEB2 during EMT, while its interactions with interleukin-11 mRNA improve Stat3 signaling to promote metastasis. The BCAR4 lncRNA, associated with breast cancer, forms the interface between SNIP1 and PNUTS transcription factors, stimulating cell migration and metastasis (Xing *et al.*, 2014).

lncRNA also take action in metastatic progression through the deregulation of chromatin. The over-expression of HOTAIR in breast cancer reprograms the epigenome through the recruitment of PRC2, imposing a mesenchymal cell phenotype that promotes breast cancer metastasis (Gupta *et al.*, 2010). Furthermore, the over-expression of HOTAIR is associated with a bad prognosis in other malignant tumors (Kogo *et al.*, 2011). SChLAP1 is a lncRNA whose expression is associated with a bad prognosis and metastatic progression (Prensner *et al.*, 2014). It promotes invasion and metastasis of prostate cancer by disturbing the tumor suppressor activity of the SWI/SNF chromatin remodeling complex (Prensner *et al.*, 2013).

The recent identification of lncRNA that suppress metastasis has opened a new perspective on the link between the tumor microenvironment and the modulation of non-coding RNA associated with metastatic progression. NKILA is a lncRNA induced by the nuclear factor kB (NFkB) in response to inflammatory signaling. It establishes a negative feedback regulation loop that suppresses NF-kB signaling by binding the NF-kB/IkB cytoplasmic complex and preventing phosphorylation of IkB and the liberation of NFkB towards the nucleus (Liu *et al.*, 2015). The weak levels of NKILA expression in human breast cancer is linked with metastatic dissemination and a bad prognosis.

LET is a lncRNA that links the hypoxic response with metastatic progression. The histone deacetylase 3, induced by hypoxia, switches off the LET promoter, inhibiting its expression and facilitating the accumulation of NF90 and hypoxic cell invasion (Yang *et al.*, 2013a).

5.3. lncRNA as diagnostic and prognostic tools and as therapeutic targets

5.3.1. *Diagnostics and prognostics*

The deregulation of lncRNA and their involvement in tumor development is an established biological reality. Since they are expressed in a highly specific manner based on cell type (normal or tumor, prostate, breast, liver, kidney, etc.), the question of their use as diagnostic and prognostic tools and also as therapeutic targets is becoming more and more common. HOTAIR, whose high level of expression in various cancers is linked to the presence of metastasis, has been characterized as a marker of a bad prognostic for breast tumors and hepatic carcinomas (Gupta *et al.*, 2010; Liu *et al.*, 2013), allowing the expansion of molecular markers for an earlier diagnosis of certain cancers.

The use of lncRNA as diagnostic tools offers several advantages due to the fact that their expression level in cells is a direct reflection of cell identity. Furthermore, in the majority of tumors, they are expressed at such levels that their presence is detected in different bodily fluids such as saliva, blood or urine (Cheetham *et al.*, 2013). For example, PCA3 is an antisense RNA expressed in prostate tumors and is a good diagnostic tool. It is measured in patients' urine and allows the confirmation of a biopsy analysis or can provide an earlier diagnosis (Foj *et al.*, 2014). Two lncRNA, HOTAIR and NEAR1, are over-expressed in oral squamous carcinomas while MEG3 is under-expressed in the same tumors. These lncRNA can therefore be directly measured in the saliva of patients and used as markers of tumors. Furthermore, HOTAIR is the only one that is over-expressed in metastatic patients (Tang *et al.*, 2013).

5.3.2. *Therapeutic targets*

lncRNA are very promising therapeutic targets due to their high specificity. Therapeutic strategies that target RNA are currently in development. Pathways are becoming clear, such as ribonucleases for example. These small proteins can enter cells through endocytosis and degrade RNA to set off cell death. However, this approach targets all RNA and can affect normal cells. Strategies based on specific RNA targeting have been tested. A plasmid carrying a gene from sub-unit A of the diphtheria toxin and under the control of the H19 gene promoter (therefore itself over-expressed in tumors) is injected into tumors and leads to a significant production of the diphtheria toxin, specifically in the tumors. This protocol has produced a significant decrease in the size of tumors during clinical trials (Gutschner and Diederichs, 2012).

Synthetic antisense RNA molecules can regulate the expression level of target RNA (regulation of transcription, of RNA stability, of their maturation and eventual translation), but can also interfere with potential RNA/protein interactions and inactivate their function. They are therefore the tool of choice.

The chimeric antisense BCL2/IGH RNA, responsible for the over-expression of BCL2 and an extremely important factor in follicular lymphomas, is an optimal target in oligodeoxynucleotide (ODN) therapy. Indeed, cells treated with ODNs targeting BCL-2/IGH have shown a large decrease in cell proliferation and sudden cell death. This therapy only targets patients with the chimeric antisense BCL2/IGH transcript, as a consequence of a chromosomal rearrangement, and has no effect on patients with a normal BCL-2 gene (Ziegler *et al.*, 1997).

During a clinical trial, an anticancer agent (Gemcitabine) administered to patients produced a cytotoxic synergy with Cisplatine. It was suggested that it was the nucleotide excision repair (NER) process that was responsible for this synergy (Yang *et al.*, 2000).

Other studies targeting various diseases have shown that the inhibition and degradation of NATs can increase the level of certain RNAs used for therapeutic purposes. The inhibition of the antisense BDNF (brain-derived neurotrophic factor) gene with molecules targeting the sense and antisense overlapping region causes an increase in the expression level of the BDNF RNA and its protein, ending in *in vivo* and *in vitro* neuronal growth and development. The same study demonstrated the effect of antisense RNA inhibition on the over-expression of two other genes: GDNF (glial-derived neurotrophic factor) and EPHB2 (ephrin receptor B2), proving that is was indeed a generalized phenomenon. These molecules, called AntagoNATs, are designed to target regions of overlap between sense and antisense to inhibit the antisense (Modarresi *et al.*, 2012). This phenomenon opens up the option of targeting antisense RNA that suppress tumor suppressor genes, such as lincRNA-p21 or ANRIL for example.

Concluding Perspectives

In conclusion, the hidden face of the genome today represents a veritable *terra incognita* whose (non-coding) genes express themselves at different key stages of a cell's life. Their role remains an enigma but the development of sequencing technologies has revealed their incredibly wide diversity. Efforts to elaborate exhaustive catalogs are far from over and many laboratories are committed to exploring the various facets. The involvement of RNA in development and in cell identity for diseases such as cancer has contributed largely to the popularity of their functional characterization. Currently, there is no feasible predictive method to understand their function, but their use in diagnostics, prognostics or as therapeutic targets is developing rapidly.

For cancer, the structural approaches used to evaluate the functions of lncRNA have revealed that the same SNP can modify the local structure of RNA in pertinent functional sites involved in binding microRNA or proteins (Wan *et al.*, 2014). Initial structural analyses of lncRNA in the cancer transcriptome revealed the major effects of SNP or somatic mutations in lncRNA associated with cancer (Chu *et al.*, 2015; Li *et al.*, 2015b; Spitale *et al.*, 2015). Furthermore, the view that certain lncRNA can code small and translated open reading frames (Anderson *et al.*, 2015; Ingolia *et al.*, 2009), and that coding RNA can have a functional role independent of its translation (Li *et al.*, 2015d) suggests that the distinction between mRNA and lncRNA may be much less important than initially thought.

Cataloging of the physiological role of lncRNA associated with cancers requires an evolution in its investigative and lncRNA labeling approach, as well as in its molecular or cellular characterization in animal genetic cancer models (Sauvageau *et al.*, 2013), as recently suggested (Li and Chang, 2014). Organoid or single cell models are particularly critical for the elucidation of the emerging physiological roles of non-coding RNA in intracellular signaling, inflammation, angiogenesis and immune modulation, all central factors in the cooperation of cancer cells and stromal cells necessary to support the growth and development of a tumor (Bernard *et al.*, 2012; Lu *et al.*, 2016; Michalik *et al.*, 2014; Satpathy and Chang, 2015; Yuan *et al.*, 2012).

Bibliography

Abdelmohsen, K., Panda, A., Kang, M.J., Xu, J., Selimyan, R., Yoon, J.H., Martindale, J.L., De, S., Wood, W.H., 3rd, Becker, K.G., *et al.* (2013). Senescence-associated lncRNAs: senescence-associated long noncoding RNAs. Aging cell *12*, 890-900.

Adams, M.D., Celniker, S.E., Holt, R.A., Evans, C.A., Gocayne, J.D., Amanatides, P.G., Scherer, S.E., Li, P.W., Hoskins, R.A., Galle, R.F., *et al.* (2000). The genome sequence of Drosophila melanogaster. Science *287*, 2185-2195.

Agarwal, N., and Ansari, A. (2016). Enhancement of Transcription by a Splicing-Competent Intron Is Dependent on Promoter Directionality. PLoS Genet *12*, e1006047.

Alam, T., Medvedeva, Y.A., Jia, H., Brown, J.B., Lipovich, L., and Bajic, V.B. (2014). Promoter analysis reveals globally differential regulation of human long non-coding RNA and protein-coding genes. PLoS One *9*, e109443.

Ameres, S.L., and Zamore, P.D. (2013). Diversifying microRNA sequence and function. Nature reviews Molecular cell biology *14*, 475-488.

An, Y., Furber, K.L., and Ji, S. (2017). Pseudogenes regulate parental gene expression via ceRNA network. Journal of cellular and molecular medicine *21*, 185-192.

Anandakumar, S., Vijayakumar, S., Arumugam, N., and Gromiha, M.M. (2015). Mammalian Mitochondrial ncRNA Database. Bioinformation *11*, 512-513.

Anderson, D.M., Anderson, K.M., Chang, C.L., Makarewich, C.A., Nelson, B.R., McAnally, J.R., Kasaragod, P., Shelton, J.M., Liou, J., Bassel-Duby, R., *et al.* (2015). A micropeptide encoded by a putative long noncoding RNA regulates muscle performance. Cell *160*, 595-606.

Andrews, S.J., and Rothnagel, J.A. (2014). Emerging evidence for functional peptides encoded by short open reading frames. Nat Rev Genet *15*, 193-204.

Antequera, F., and Bird, A. (1993). Number of CpG islands and genes in human and mouse. Proc Natl Acad Sci U S A *90*, 11995-11999.

Arab, K., Park, Y.J., Lindroth, A.M., Schafer, A., Oakes, C., Weichenhan, D., Lukanova, A., Lundin, E., Risch, A., Meister, M., *et al.* (2014). Long noncoding RNA TARID directs demethylation and activation of the tumor suppressor TCF21 via GADD45A. Mol Cell *55*, 604-614.

Ariel, F., Romero-Barrios, N., Jegu, T., Benhamed, M., and Crespi, M. (2015). Battles and hijacks: noncoding transcription in plants. Trends in plant science *20*, 362-371.

Avery, O.T., Macleod, C.M., and McCarty, M. (1944). Studies on the Chemical Nature of the Substance Inducing Transformation of Pneumococcal Types : Induction of Transformation by a Desoxyribonucleic Acid Fraction Isolated from Pneumococcus Type Iii. The Journal of experimental medicine *79*, 137-158.

Ayupe, A.C., Tahira, A.C., Camargo, L., Beckedorff, F.C., Verjovski-Almeida, S., and Reis, E.M. (2015). Global analysis of biogenesis, stability and sub-cellular localization of lncRNAs mapping to intragenic regions of the human genome. RNA biology *12*, 877-892.

Azzalin, C.M., Reichenbach, P., Khoriauli, L., Giulotto, E., and Lingner, J. (2007). Telomeric repeat containing RNA and RNA surveillance factors at mammalian chromosome ends. Science *318*, 798-801.

Bae, B.I., Tietjen, I., Atabay, K.D., Evrony, G.D., Johnson, M.B., Asare, E., Wang, P.P., Murayama, A.Y., Im, K., Lisgo, S.N., *et al.* (2014). Evolutionarily dynamic alternative splicing of GPR56 regulates regional cerebral cortical patterning. Science *343*, 764-768.

Banfai, B., Jia, H., Khatun, J., Wood, E., Risk, B., Gundling, W.E., Jr., Kundaje, A., Gunawardena, H.P., Yu, Y., Xie, L., *et al.* (2012). Long noncoding RNAs are rarely translated in two human cell lines. Genome Res *22*, 1646-1657.

Barber, B.A., and Rastegar, M. (2010). Epigenetic control of Hox genes during neurogenesis, development, and disease. Annals of anatomy = Anatomischer Anzeiger : official organ of the Anatomische Gesellschaft *192*, 261-274.

Barlow, D.P., Stoger, R., Herrmann, B.G., Saito, K., and Schweifer, N. (1991). The mouse insulin-like growth factor type-2 receptor is imprinted and closely linked to the Tme locus. Nature *349*, 84-87.

Bartolomei, M.S., Zemel, S., and Tilghman, S.M. (1991). Parental imprinting of the mouse H19 gene. Nature *351*, 153-155.

Bejerano, G., Pheasant, M., Makunin, I., Stephen, S., Kent, W.J., Mattick, J.S., and Haussler, D. (2004). Ultraconserved elements in the human genome. Science *304*, 1321-1325.

Beltran, M., Puig, I., Pena, C., Garcia, J.M., Alvarez, A.B., Pena, R., Bonilla, F., and de Herreros, A.G. (2008). A natural antisense transcript regulates Zeb2/Sip1 gene expression during Snail1-induced epithelial-mesenchymal transition. Genes Dev *22*, 756-769.

Beniaminov, A., Westhof, E., and Krol, A. (2008). Distinctive structures between chimpanzee and human in a brain noncoding RNA. Rna *14*, 1270-1275.

Berk, A.J. (2016). Discovery of RNA splicing and genes in pieces. Proc Natl Acad Sci U S A *113*, 801-805.

Bernard, D., Prasanth, K.V., Tripathi, V., Colasse, S., Nakamura, T., Xuan, Z., Zhang, M.Q., Sedel, F., Jourdren, L., Coulpier, F., *et al.* (2010). A long nuclear-retained non-coding RNA regulates synaptogenesis by modulating gene expression. Embo J *29*, 3082-3093.

Bernard, J.J., Cowing-Zitron, C., Nakatsuji, T., Muehleisen, B., Muto, J., Borkowski, A.W., Martinez, L., Greidinger, E.L., Yu, B.D., and Gallo, R.L. (2012). Ultraviolet radiation damages self noncoding RNA and is detected by TLR3. Nature medicine *18*, 1286-1290.

Bernhardt, H.S. (2012). The RNA world hypothesis: the worst theory of the early evolution of life (except for all the others)(a). Biology direct *7*, 23.

Beroukhim, R., Mermel, C.H., Porter, D., Wei, G., Raychaudhuri, S., Donovan, J., Barretina, J., Boehm, J.S., Dobson, J., Urashima, M., *et al.* (2010). The landscape of somatic copy-number alteration across human cancers. Nature *463*, 899-905.

Berretta, J., and Morillon, A. (2009). Pervasive transcription constitutes a new level of eukaryotic genome regulation. EMBO Rep *10*, 973-982.

Berretta, J., Pinskaya, M., and Morillon, A. (2008). A cryptic unstable transcript mediates transcriptional trans-silencing of the Ty1 retrotransposon in S. cerevisiae. Genes Dev *22*, 615-626.

Bertani, S., Sauer, S., Bolotin, E., and Sauer, F. (2011). The noncoding RNA Mistral activates Hoxa6 and Hoxa7 expression and stem cell differentiation by recruiting MLL1 to chromatin. Mol Cell *43*, 1040-1046.

Bhan, A., and Mandal, S.S. (2015). LncRNA HOTAIR: A master regulator of chromatin dynamics and cancer. Biochimica et biophysica acta *1856*, 151-164.

Bianchessi, V., Badi, I., Bertolotti, M., Nigro, P., D'Alessandra, Y., Capogrossi, M.C., Zanobini, M., Pompilio, G., Raucci, A., and Lauri, A. (2015). The mitochondrial lncRNA ASncmtRNA-2 is induced in aging and replicative senescence in Endothelial Cells. Journal of molecular and cellular cardiology *81*, 62-70.

Bierhoff, H., Dammert, M.A., Brocks, D., Dambacher, S., Schotta, G., and Grummt, I. (2014). Quiescence-induced LncRNAs trigger H4K20 trimethylation and transcriptional silencing. Mol Cell *54*, 675-682.

Bird, C.P., Stranger, B.E., Liu, M., Thomas, D.J., Ingle, C.E., Beazley, C., Miller, W., Hurles, M.E., and Dermitzakis, E.T. (2007). Fast-evolving noncoding sequences in the human genome. Genome Biol *8*, R118.

Birney, E., Stamatoyannopoulos, J.A., Dutta, A., Guigo, R., Gingeras, T.R., Margulies, E.H., Weng, Z., Snyder, M., Dermitzakis, E.T., Thurman, R.E., *et al.* (2007). Identification and analysis of functional elements in 1% of the human genome by the ENCODE pilot project. Nature *447*, 799-816.

Blower, M.D. (2016). Centromeric Transcription Regulates Aurora-B Localization and Activation. Cell reports *15*, 1624-1633.

Bond, A.M., Vangompel, M.J., Sametsky, E.A., Clark, M.F., Savage, J.C., Disterhoft, J.F., and Kohtz, J.D. (2009). Balanced gene regulation by an embryonic brain ncRNA is critical for adult hippocampal GABA circuitry. Nature neuroscience *12*, 1020-1027.

Borsani, G., Tonlorenzi, R., Simmler, M.C., Dandolo, L., Arnaud, D., Capra, V., Grompe, M., Pizzuti, A., Muzny, D., Lawrence, C., *et al.* (1991). Characterization of a murine gene expressed from the inactive X chromosome. Nature *351*, 325-329.

Bottcher, R., Hoogland, A.M., Dits, N., Verhoef, E.I., Kweldam, C., Waranecki, P., Bangma, C.H., van Leenders, G.J., and Jenster, G. (2015). Novel long non-coding RNAs are specific diagnostic and prognostic markers for prostate cancer. Oncotarget *6*, 4036-4050.

Brannan, C.I., Dees, E.C., Ingram, R.S., and Tilghman, S.M. (1990). The product of the H19 gene may function as an RNA. Mol Cell Biol *10*, 28-36.

Broadbent, K.M., Broadbent, J.C., Ribacke, U., Wirth, D., Rinn, J.L., and Sabeti, P.C. (2015). Strand-specific RNA sequencing in Plasmodium falciparum malaria identifies developmentally regulated long non-coding RNA and circular RNA. BMC genomics *16*, 454.

Brockdorff, N., Ashworth, A., Kay, G.F., Cooper, P., Smith, S., McCabe, V.M., Norris, D.P., Penny, G.D., Patel, D., and Rastan, S. (1991). Conservation of position and exclusive expression of mouse Xist from the inactive X chromosome. Nature *351*, 329-331.

Brown, C.J., Ballabio, A., Rupert, J.L., Lafreniere, R.G., Grompe, M., Tonlorenzi, R., and Willard, H.F. (1991). A gene from the region of the human X inactivation centre is expressed exclusively from the inactive X chromosome. Nature *349*, 38-44.

Brown, C.J., and Willard, H.F. (1994). The human X-inactivation centre is not required for maintenance of X-chromosome inactivation. Nature *368*, 154-156.

Brunner, A.L., Beck, A.H., Edris, B., Sweeney, R.T., Zhu, S.X., Li, R., Montgomery, K., Varma, S., Gilks, T., Guo, X., *et al.* (2012). Transcriptional profiling of lncRNAs and novel transcribed regions across a diverse panel of archived human cancers. Genome Biol *13*, R75.

Burzio, V.A., Villota, C., Villegas, J., Landerer, E., Boccardo, E., Villa, L.L., Martinez, R., Lopez, C., Gaete, F., Toro, V., *et al.* (2009). Expression of a family of noncoding mitochondrial RNAs distinguishes normal from cancer cells. Proc Natl Acad Sci U S A *106*, 9430-9434.

Bussemakers, M.J., van Bokhoven, A., Verhaegh, G.W., Smit, F.P., Karthaus, H.F., Schalken, J.A., Debruyne, F.M., Ru, N., and Isaacs, W.B. (1999). DD3: a new prostate-specific gene, highly overexpressed in prostate cancer. Cancer Res *59*, 5975-5979.

Butcher, S.E. (2009). The spliceosome as ribozyme hypothesis takes a second step. Proc Natl Acad Sci U S A *106*, 12211-12212.

Cabianca, D.S., Casa, V., Bodega, B., Xynos, A., Ginelli, E., Tanaka, Y., and Gabellini, D. (2012). A long ncRNA links copy number variation to a polycomb/trithorax epigenetic switch in FSHD muscular dystrophy. Cell *149*, 819-831.

Cabili, M.N., Dunagin, M.C., McClanahan, P.D., Biaesch, A., Padovan-Merhar, O., Regev, A., Rinn, J.L., and Raj, A. (2015). Localization and abundance analysis of human lncRNAs at single-cell and single-molecule resolution. Genome Biol *16*, 20.

Cajigas, I., Leib, D.E., Cochrane, J., Luo, H., Swyter, K.R., Chen, S., Clark, B.S., Thompson, J., Yates, J.R., 3rd, Kingston, R.E., *et al.* (2015). Evf2 lncRNA/BRG1/DLX1 interactions reveal RNA-dependent inhibition of chromatin remodeling. Development *142*, 2641-2652.

Calin, G.A., Liu, C.G., Ferracin, M., Hyslop, T., Spizzo, R., Sevignani, C., Fabbri, M., Cimmino, A., Lee, E.J., Wojcik, S.E., *et al.* (2007). Ultraconserved regions encoding ncRNAs are altered in human leukemias and carcinomas. Cancer cell *12*, 215-229.

Camblong, J., Beyrouthy, N., Guffanti, E., Schlaepfer, G., Steinmetz, L.M., and Stutz, F. (2009). Trans-acting antisense RNAs mediate transcriptional gene cosuppression in S. cerevisiae. Genes Dev *23*, 1534-1545.

Carlile, M., Swan, D., Jackson, K., Preston-Fayers, K., Ballester, B., Flicek, P., and Werner, A. (2009). Strand selective generation of endo-siRNAs from the Na/phosphate transporter gene Slc34a1 in murine tissues. Nucleic Acids Res *37*, 2274-2282.

Carninci, P., Kasukawa, T., Katayama, S., Gough, J., Frith, M.C., Maeda, N., Oyama, R., Ravasi, T., Lenhard, B., Wells, C., *et al.* (2005). The transcriptional landscape of the mammalian genome. Science *309*, 1559-1563.

Carrieri, C., Cimatti, L., Biagioli, M., Beugnet, A., Zucchelli, S., Fedele, S., Pesce, E., Ferrer, I., Collavin, L., Santoro, C., *et al.* (2012). Long non-coding antisense RNA controls Uchl1 translation through an embedded SINEB2 repeat. Nature *491*, 454-457.

Castel, S.E., and Martienssen, R.A. (2013). RNA interference in the nucleus: roles for small RNAs in transcription, epigenetics and beyond. Nat Rev Genet *14*, 100-112.

Cech, T.R. (2000). Structural biology. The ribosome is a ribozyme. Science *289*, 878-879.

Cesana, M., Cacchiarelli, D., Legnini, I., Santini, T., Sthandier, O., Chinappi, M., Tramontano, A., and Bozzoni, I. (2011). A long noncoding RNA controls muscle differentiation by functioning as a competing endogenous RNA. Cell *147*, 358-369.

Chan, F.L., Marshall, O.J., Saffery, R., Kim, B.W., Earle, E., Choo, K.H., and Wong, L.H. (2012). Active transcription and essential role of RNA polymerase II at the centromere during mitosis. Proc Natl Acad Sci U S A *109*, 1979-1984.

Cheetham, S.W., Gruhl, F., Mattick, J.S., and Dinger, M.E. (2013). Long noncoding RNAs and the genetics of cancer. British journal of cancer *108*, 2419-2425.

Chen, L.L. (2016). Linking Long Noncoding RNA Localization and Function. Trends Biochem Sci *41*, 761-772.

Choudhry, H., Harris, A.L., and McIntyre, A. (2016). The tumour hypoxia induced non-coding transcriptome. Molecular aspects of medicine *47-48*, 35-53.

Chu, C., Spitale, R.C., and Chang, H.Y. (2015). Technologies to probe functions and mechanisms of long noncoding RNAs. Nat Struct Mol Biol *22*, 29-35.

Chujo, T., Yamazaki, T., and Hirose, T. (2016). Architectural RNAs (arcRNAs): A class of long noncoding RNAs that function as the scaffold of nuclear bodies. Biochimica et biophysica acta *1859*, 139-146.

Chureau, C., Prissette, M., Bourdet, A., Barbe, V., Cattolico, L., Jones, L., Eggen, A., Avner, P., and Duret, L. (2002). Comparative sequence analysis of the X-inactivation center region in mouse, human, and bovine. Genome Res *12*, 894-908.

Clark, M.B., Choudhary, A., Smith, M.A., Taft, R.J., and Mattick, J.S. (2013). The dark matter rises: the expanding world of regulatory RNAs. Essays in biochemistry *54*, 1-16.

Clark, M.B., Johnston, R.L., Inostroza-Ponta, M., Fox, A.H., Fortini, E., Moscato, P., Dinger, M.E., and Mattick, J.S. (2012). Genome-wide analysis of long noncoding RNA stability. Genome Res *22*, 885-898.

Clemson, C.M., Hutchinson, J.N., Sara, S.A., Ensminger, A.W., Fox, A.H., Chess, A., and Lawrence, J.B. (2009). An architectural role for a nuclear noncoding RNA: NEAT1 RNA is essential for the structure of paraspeckles. Mol Cell *33*, 717-726.

Cobb, M. (2015). Who discovered messenger RNA? Current biology : CB *25*, R526-532.

Colin, J., Libri, D., and Porrua, O. (2011). Cryptic transcription and early termination in the control of gene expression. Genetics research international *2011*, 653494.

Consortium, E.p. (2012). An integrated encyclopedia of DNA elements in the human genome. Nature *489*, 57-74.

Consortium, H.G.s. (2004). Finishing the euchromatic sequence of the human genome. Nature *431*, 931-945.

Corcoran, C.P., Podkaminski, D., Papenfort, K., Urban, J.H., Hinton, J.C., and Vogel, J. (2012). Superfolder GFP reporters validate diverse new mRNA targets of the classic porin regulator, MicF RNA. Molecular microbiology *84*, 428-445.

Crick, F.H. (1968). The origin of the genetic code. Journal of molecular biology *38*, 367-379.

Dahm, R. (2005). Friedrich Miescher and the discovery of DNA. Developmental biology *278*, 274-288.

David, L., Huber, W., Granovskaia, M., Toedling, J., Palm, C.J., Bofkin, L., Jones, T., Davis, R.W., and Steinmetz, L.M. (2006). A high-resolution map of transcription in the yeast genome. Proc Natl Acad Sci U S A *103*, 5320-5325.

De Andres-Pablo, A., Morillon, A., and Wery, M. (2016). LncRNAs, lost in translation or licence to regulate? Current genetics.

Delihas, N. (2015). Discovery and characterization of the first non-coding RNA that regulates gene expression, micF RNA: A historical perspective. World journal of biological chemistry *6*, 272-280.

Derrien, T., Johnson, R., Bussotti, G., Tanzer, A., Djebali, S., Tilgner, H., Guernec, G., Martin, D., Merkel, A., Knowles, D.G., *et al.* (2012). The GENCODE v7 catalog of human long noncoding RNAs: Analysis of their gene structure, evolution, and expression. Genome Res *22*, 1775-1789.

Dey, B.K., Pfeifer, K., and Dutta, A. (2014). The H19 long noncoding RNA gives rise to microRNAs miR-675-3p and miR-675-5p to promote skeletal muscle differentiation and regeneration. Genes Dev *28*, 491-501.

Dhir, A., Dhir, S., Proudfoot, N.J., and Jopling, C.L. (2015). Microprocessor mediates transcriptional termination of long noncoding RNA transcripts hosting microRNAs. Nat Struct Mol Biol *22*, 319-327.

Dimitrova, N., Zamudio, J.R., Jong, R.M., Soukup, D., Resnick, R., Sarma, K., Ward, A.J., Raj, A., Lee, J.T., Sharp, P.A., *et al.* (2014). LincRNA-p21 activates p21 in cis to promote Polycomb target gene expression and to enforce the G1/S checkpoint. Mol Cell *54*, 777-790.

Dinger, M., Amaral, P., Mercer, T., Pang, K., Bruce, S., Gardiner, B., Askarian-Amiri, M., Ru, K., Soldà, G., Simons, C., *et al.* (2008). Long noncoding RNAs in mouse embryonic stem cell pluripotency and differentiation. Genome Res *18*, 1433-1445.

Dinger, M., Amaral, P., Mercer, T., and Mattick, J.S. (2009). Pervasive transcription of the eukaryotic genome: functional indices and conceptual implications. Briefings in functional genomics & proteomics *8*, 407-423.

Djebali, S., Davis, C.A., Merkel, A., Dobin, A., Lassmann, T., Mortazavi, A., Tanzer, A., Lagarde, J., Lin, W., Schlesinger, F., *et al.* (2012). Landscape of transcription in human cells. Nature *489*, 101-108.

Doan, R.N., Bae, B.I., Cubelos, B., Chang, C., Hossain, A.A., Al-Saad, S., Mukaddes, N.M., Oner, O., Al-Saffar, M., Balkhy, S., *et al.* (2016). Mutations in Human Accelerated Regions Disrupt Cognition and Social Behavior. Cell *167*, 341-354 e312.

Dorji, T., Monti, V., Fellegara, G., Gabba, S., Grazioli, V., Repetti, E., Marcialis, C., Peluso, S., Di Ruzza, D., Neri, F., *et al.* (2015). Gain of hTERC: a genetic marker of malignancy in oral potentially malignant lesions. Human pathology *46*, 1275-1281.

Dorus, S., Vallender, E.J., Evans, P.D., Anderson, J.R., Gilbert, S.L., Mahowald, M., Wyckoff, G.J., Malcom, C.M., and Lahn, B.T. (2004). Accelerated evolution of nervous system genes in the origin of Homo sapiens. Cell *119*, 1027-1040.

Du, Z., Fei, T., Verhaak, R.G., Su, Z., Zhang, Y., Brown, M., Chen, Y., and Liu, X.S. (2013). Integrative genomic analyses reveal clinically relevant long noncoding RNAs in human cancer. Nat Struct Mol Biol *20*, 908-913.

Dunham, I., Shimizu, N., Roe, B.A., Chissoe, S., Hunt, A.R., Collins, J.E., Bruskiewich, R., Beare, D.M., Clamp, M., Smink, L.J., *et al.* (1999). The DNA sequence of human chromosome 22. Nature *402*, 489-495.

Duret, L., Chureau, C., Samain, S., Weissenbach, J., and Avner, P. (2006). The Xist RNA gene evolved in eutherians by pseudogenization of a protein-coding gene. Science *312*, 1653-1655.

Durruthy-Durruthy, J., Sebastiano, V., Wossidlo, M., Cepeda, D., Cui, J., Grow, E.J., Davila, J., Mall, M., Wong, W.H., Wysocka, J., *et al.* (2016). The primate-specific noncoding RNA HPAT5 regulates pluripotency during human preimplantation development and nuclear reprogramming. Nat Genet *48*, 44-52.

Eades, G., Zhang, Y.S., Li, Q.L., Xia, J.X., Yao, Y., and Zhou, Q. (2014). Long non-coding RNAs in stem cells and cancer. World journal of clinical oncology *5*, 134-141.

Eichhorn, S.W., Subtelny, A.O., Kronja, I., Kwasnieski, J.C., Orr-Weaver, T.L., and Bartel, D.P. (2016). mRNA poly(A)-tail changes specified by deadenylation broadly reshape translation in Drosophila oocytes and early embryos. eLife *5*.

Engreitz, J.M., Pandya-Jones, A., McDonel, P., Shishkin, A., Sirokman, K., Surka, C., Kadri, S., Xing, J., Goren, A., Lander, E.S., *et al.* (2013). The Xist lncRNA exploits three-dimensional genome architecture to spread across the X chromosome. Science *341*, 1237973.

Enuka, Y., Lauriola, M., Feldman, M.E., Sas-Chen, A., Ulitsky, I., and Yarden, Y. (2016). Circular RNAs are long-lived and display only minimal early alterations in response to a growth factor. Nucleic Acids Res *44*, 1370-1383.

Espinoza, C.A., Goodrich, J.A., and Kugel, J.F. (2007). Characterization of the structure, function, and mechanism of B2 RNA, an ncRNA repressor of RNA polymerase II transcription. Rna *13*, 583-596.

Faghihi, M.A., Modarresi, F., Khalil, A.M., Wood, D.E., Sahagan, B.G., Morgan, T.E., Finch, C.E., St Laurent, G., 3rd, Kenny, P.J., and Wahlestedt, C. (2008). Expression of a noncoding RNA is elevated in Alzheimer's disease and drives rapid feed-forward regulation of beta-secretase. Nature medicine *14*, 723-730.

Fassan, M., Dall'Olmo, L., Galasso, M., Braconi, C., Pizzi, M., Realdon, S., Volinia, S., Valeri, N., Gasparini, P., Baffa, R., *et al.* (2014). Transcribed ultraconserved noncoding RNAs (T-UCR) are involved in Barrett's esophagus carcinogenesis. Oncotarget *5*, 7162-7171.

Feng, J., Bi, C., Clark, B.S., Mady, R., Shah, P., and Kohtz, J.D. (2006). The Evf-2 noncoding RNA is transcribed from the Dlx-5/6 ultraconserved region and functions as a Dlx-2 transcriptional coactivator. Genes Dev *20*, 1470-1484.

Feng, J., Funk, W.D., Wang, S.S., Weinrich, S.L., Avilion, A.A., Chiu, C.P., Adams, R.R., Chang, E., Allsopp, R.C., Yu, J., *et al.* (1995). The RNA component of human telomerase. Science *269*, 1236-1241.

Ferdin, J., Nishida, N., Wu, X., Nicoloso, M.S., Shah, M.Y., Devlin, C., Ling, H., Shimizu, M., Kumar, K., Cortez, M.A., *et al.* (2013). HINCUTs in cancer: hypoxia-induced noncoding ultraconserved transcripts. Cell death and differentiation *20*, 1675-1687.

Feuerhahn, S., Iglesias, N., Panza, A., Porro, A., and Lingner, J. (2010). TERRA biogenesis, turnover and implications for function. FEBS letters *584*, 3812-3818.

Fire, A., Xu, S., Montgomery, M.K., Kostas, S.A., Driver, S.E., and Mello, C.C. (1998). Potent and specific genetic interference by double-stranded RNA in Caenorhabditis elegans. Nature *391*, 806-811.

Fitzpatrick, G.V., Soloway, P.D., and Higgins, M.J. (2002). Regional loss of imprinting and growth deficiency in mice with a targeted deletion of KvDMR1. Nat Genet *32*, 426-431.

Fleischmann, R.D., Adams, M.D., White, O., Clayton, R.A., Kirkness, E.F., Kerlavage, A.R., Bult, C.J., Tomb, J.F., Dougherty, B.A., Merrick, J.M., *et al.* (1995). Whole-genome random sequencing and assembly of Haemophilus influenzae Rd. Science *269*, 496-512.

Flockhart, R.J., Webster, D.E., Qu, K., Mascarenhas, N., Kovalski, J., Kretz, M., and Khavari, P.A. (2012). BRAFV600E remodels the melanocyte transcriptome and induces BANCR to regulate melanoma cell migration. Genome Res *22*, 1006-1014.

Flynn, R.A., Almada, A.E., Zamudio, J.R., and Sharp, P.A. (2011a). Antisense RNA polymerase II divergent transcripts are P-TEFb dependent and substrates for the RNA exosome. Proc Natl Acad Sci U S A *108*, 10460-10465.

Flynn, R.A., and Chang, H.Y. (2014). Long noncoding RNAs in cell-fate programming and reprogramming. Cell stem cell *14*, 752-761.

Flynn, R.L., Centore, R.C., O'Sullivan, R.J., Rai, R., Tse, A., Songyang, Z., Chang, S., Karlseder, J., and Zou, L. (2011b). TERRA and hnRNPA1 orchestrate an RPA-to-POT1 switch on telomeric single-stranded DNA. Nature *471*, 532-536.

Foj, L., Mila, M., Mengual, L., Luque, P., Alcaraz, A., Jimenez, W., and Filella, X. (2014). Real-time PCR PCA3 assay is a useful test measured in urine to improve prostate cancer detection. Clinica chimica acta; international journal of clinical chemistry *435*, 53-58.

Fox, M.J., Gao, H., Smith-Kinnaman, W.R., Liu, Y., and Mosley, A.L. (2015). The exosome component Rrp6 is required for RNA polymerase II termination at specific targets of the Nrd1-Nab3 pathway. PLoS Genet *11*, e1004999.

Ganesh, S., and Svoboda, P. (2016). Retrotransposon-associated long non-coding RNAs in mice and men. Pflugers Archiv : European journal of physiology *468*, 1049-1060.

Gardner, E.J., Nizami, Z.F., Talbot, C.C., Jr., and Gall, J.G. (2012). Stable intronic sequence RNA (sisRNA), a new class of noncoding RNA from the oocyte nucleus of Xenopus tropicalis. Genes Dev *26*, 2550-2559.

Gendrel, A.V., and Heard, E. (2014). Noncoding RNAs and epigenetic mechanisms during X-chromosome inactivation. Annual review of cell and developmental biology *30*, 561-580.

Giannakakis, A., Zhang, J., Jenjaroenpun, P., Nama, S., Zainolabidin, N., Aau, M.Y., Yarmishyn, A.A., Vaz, C., Ivshina, A.V., Grinchuk, O.V., *et al.* (2015). Contrasting expression patterns of coding and noncoding parts of the human genome upon oxidative stress. Scientific reports *5*, 9737.

Gibb, E.A., Brown, C.J., and Lam, W.L. (2011). The functional role of long non-coding RNA in human carcinomas. Molecular cancer *10*, 38.

Giovarelli, M., Bucci, G., Ramos, A., Bordo, D., Wilusz, C.J., Chen, C.Y., Puppo, M., Briata, P., and Gherzi, R. (2014). H19 long noncoding RNA controls the mRNA decay promoting function of KSRP. Proc Natl Acad Sci U S A *111*, E5023-5028.

Goffeau, A., Barrell, B.G., Bussey, H., Davis, R.W., Dujon, B., Feldmann, H., Galibert, F., Hoheisel, J.D., Jacq, C., Johnston, M., *et al.* (1996). Life with 6000 genes. Science *274*, 546, 563-547.

Gomez, J.A., Wapinski, O.L., Yang, Y.W., Bureau, J.F., Gopinath, S., Monack, D.M., Chang, H.Y., Brahic, M., and Kirkegaard, K. (2013). The NeST long ncRNA controls microbial susceptibility and epigenetic activation of the interferon-gamma locus. Cell *152*, 743-754.

Gonzalez, I., Munita, R., Agirre, E., Dittmer, T.A., Gysling, K., Misteli, T., and Luco, R.F. (2015). A lncRNA regulates alternative splicing via establishment of a splicing-specific chromatin signature. Nat Struct Mol Biol *22*, 370-376.

Goodman, A.J., Daugharthy, E.R., and Kim, J. (2013). Pervasive antisense transcription is evolutionarily conserved in budding yeast. Molecular biology and evolution *30*, 409-421.

Greenwood, J., and Cooper, J.P. (2012). Non-coding telomeric and subtelomeric transcripts are differentially regulated by telomeric and heterochromatin assembly factors in fission yeast. Nucleic Acids Res *40*, 2956-2963.

Grote, P., Wittler, L., Hendrix, D., Koch, F., Wahrisch, S., Beisaw, A., Macura, K., Blass, G., Kellis, M., Werber, M., *et al.* (2013). The tissue-specific lncRNA Fendrr is an essential regulator of heart and body wall development in the mouse. Developmental cell *24*, 206-214.

Guenzl, P.M., and Barlow, D.P. (2012). Macro lncRNAs: a new layer of cis-regulatory information in the mammalian genome. RNA biology *9*, 731-741.

Guerrier-Takada, C., Gardiner, K., Marsh, T., Pace, N., and Altman, S. (1983). The RNA moiety of ribonuclease P is the catalytic subunit of the enzyme. Cell *35*, 849-857.

Guo, X., Gao, L., Wang, Y., Chiu, D.K., Wang, T., and Deng, Y. (2016). Advances in long noncoding RNAs: identification, structure prediction and function annotation. Briefings in functional genomics *15*, 38-46.

Gupta, R.A., Shah, N., Wang, K.C., Kim, J., Horlings, H.M., Wong, D.J., Tsai, M.C., Hung, T., Argani, P., Rinn, J.L., *et al.* (2010). Long non-coding RNA HOTAIR reprograms chromatin state to promote cancer metastasis. Nature *464*, 1071-1076.

Gutschner, T., and Diederichs, S. (2012). The hallmarks of cancer: a long non-coding RNA point of view. RNA biology *9*, 703-719.

Gutschner, T., Hammerle, M., Eissmann, M., Hsu, J., Kim, Y., Hung, G., Revenko, A., Arun, G., Stentrup, M., Gross, M., *et al.* (2013). The Noncoding RNA MALAT1 Is a Critical Regulator of the Metastasis Phenotype of Lung Cancer Cells. Cancer Res *73*, 1180-1189.

Guttman, M., Amit, I., Garber, M., French, C., Lin, M.F., Feldser, D., Huarte, M., Zuk, O., Carey, B.W., Cassady, J.P., *et al.* (2009). Chromatin signature reveals over a thousand highly conserved large non-coding RNAs in mammals. Nature *458*, 223-227.

Guttman, M., Donaghey, J., Carey, B.W., Garber, M., Grenier, J.K., Munson, G., Young, G., Lucas, A.B., Ach, R., Bruhn, L., *et al.* (2011). lincRNAs act in the circuitry controlling pluripotency and differentiation. Nature *477*, 295-300.

Guttman, M., Garber, M., Levin, J.Z., Donaghey, J., Robinson, J., Adiconis, X., Fan, L., Koziol, M.J., Gnirke, A., Nusbaum, C., *et al.* (2010). Ab initio reconstruction of cell type-specific transcriptomes in mouse reveals the conserved multi-exonic structure of lincRNAs. Nat Biotechnol *28*, 503-510.

Ha, H., Song, J., Wang, S., Kapusta, A., Feschotte, C., Chen, K.C., and Xing, J. (2014). A comprehensive analysis of piRNAs from adult human testis and their relationship with genes and mobile elements. BMC genomics *15*, 545.

Hadjiargyrou, M., and Delihas, N. (2013). The intertwining of transposable elements and non-coding RNAs. International journal of molecular sciences *14*, 13307-13328.

Hamazaki, N., Uesaka, M., Nakashima, K., Agata, K., and Imamura, T. (2015). Gene activation-associated long noncoding RNAs function in mouse preimplantation development. Development *142*, 910-920.

Hanahan, D., and Weinberg, R.A. (2000). The hallmarks of cancer. Cell *100*, 57-70.

Hanahan, D., and Weinberg, R.A. (2011). Hallmarks of cancer: the next generation. Cell *144*, 646-674.

Hansen, T.B., Jensen, T.I., Clausen, B.H., Bramsen, J.B., Finsen, B., Damgaard, C.K., and Kjems, J. (2013). Natural RNA circles function as efficient microRNA sponges. Nature *495*, 384-388.

Hansen, T.B., Wiklund, E.D., Bramsen, J.B., Villadsen, S.B., Statham, A.L., Clark, S.J., and Kjems, J. (2011). miRNA-dependent gene silencing involving Ago2-mediated cleavage of a circular antisense RNA. Embo J *30*, 4414-4422.

Harrow, J., Frankish, A., Gonzalez, J.M., Tapanari, E., Diekhans, M., Kokocinski, F., Aken, B.L., Barrell, D., Zadissa, A., Searle, S., *et al.* (2012). GENCODE: the reference human genome annotation for The ENCODE Project. Genome Res *22*, 1760-1774.

He, L., and Hannon, G. (2004). MicroRNAs: small RNAs with a big role in gene regulation. Nat Rev Genet *5*, 522-531.

He, S., Liu, S., and Zhu, H. (2011). The sequence, structure and evolutionary features of HOTAIR in mammals. BMC evolutionary biology *11*, 102.

Heard, E., Mongelard, F., Arnaud, D., Chureau, C., Vourc'h, C., and Avner, P. (1999). Human XIST yeast artificial chromosome transgenes show partial X inactivation center function in mouse embryonic stem cells. Proc Natl Acad Sci U S A *96*, 6841-6846.

Heinen, T.J., Staubach, F., Haming, D., and Tautz, D. (2009). Emergence of a new gene from an intergenic region. Current biology : CB *19*, 1527-1531.

Hezroni, H., Koppstein, D., Schwartz, M.G., Avrutin, A., Bartel, D.P., and Ulitsky, I. (2015). Principles of long noncoding RNA evolution derived from direct comparison of transcriptomes in 17 species. Cell reports *11*, 1110-1122.

Holland, A.J., and Cleveland, D.W. (2009). Boveri revisited: chromosomal instability, aneuploidy and tumorigenesis. Nature reviews Molecular cell biology *10*, 478-487.

Housman, G., and Ulitsky, I. (2016). Methods for distinguishing between protein-coding and long noncoding RNAs and the elusive biological purpose of translation of long noncoding RNAs. Biochimica et biophysica acta *1859*, 31-40.

Hu, H.Y., He, L., and Khaitovich, P. (2014a). Deep sequencing reveals a novel class of bidirectional promoters associated with neuronal genes. BMC genomics *15*, 457.

Hu, W., Yuan, B., Flygare, J., and Lodish, H.F. (2011). Long noncoding RNA-mediated anti-apoptotic activity in murine erythroid terminal differentiation. Genes Dev *25*, 2573-2578.

Hu, X., Feng, Y., Zhang, D., Zhao, S.D., Hu, Z., Greshock, J., Zhang, Y., Yang, L., Zhong, X., Wang, L.P., *et al.* (2014b). A functional genomic approach identifies FAL1 as an oncogenic long noncoding RNA that associates with BMI1 and represses p21 expression in cancer. Cancer cell *26*, 344-357.

Hu, Y., Wang, J., Qian, J., Kong, X., Tang, J., Wang, Y., Chen, H., Hong, J., Zou, W., Chen, Y., *et al.* (2014c). Long noncoding RNA GAPLINC regulates CD44-dependent cell invasiveness and associates with poor prognosis of gastric cancer. Cancer Res *74*, 6890-6902.

Huarte, M. (2015). The emerging role of lncRNAs in cancer. Nature medicine *21*, 1253-1261.

Huarte, M., Guttman, M., Feldser, D., Garber, M., Koziol, M.J., Kenzelmann-Broz, D., Khalil, A.M., Zuk, O., Amit, I., Rabani, M., *et al.* (2010). A large intergenic noncoding RNA induced by p53 mediates global gene repression in the p53 response. Cell *142*, 409-419.

Hudson, W.H., Pickard, M.R., de Vera, I.M., Kuiper, E.G., Mourtada-Maarabouni, M., Conn, G.L., Kojetin, D.J., Williams, G.T., and Ortlund, E.A. (2014). Conserved sequence-specific lincRNA-steroid receptor interactions drive transcriptional repression and direct cell fate. Nat Commun *5*, 5395.

Hung, C.L., Wang, L.Y., Yu, Y.L., Chen, H.W., Srivastava, S., Petrovics, G., and Kung, H.J. (2014). A long noncoding RNA connects c-Myc to tumor metabolism. Proc Natl Acad Sci U S A *111*, 18697-18702.

Hung, T., Wang, Y., Lin, M.F., Koegel, A.K., Kotake, Y., Grant, G.D., Horlings, H.M., Shah, N., Umbricht, C., Wang, P., *et al.* (2011). Extensive and coordinated transcription of noncoding RNAs within cell-cycle promoters. Nat Genet *43*, 621-629.

Indrieri, A., Grimaldi, C., Zucchelli, S., Tammaro, R., Gustincich, S., and Franco, B. (2016). Synthetic long non-coding RNAs [SINEUPs] rescue defective gene expression in vivo. Scientific reports *6*, 27315.

Ingolia, N.T., Ghaemmaghami, S., Newman, J.R., and Weissman, J.S. (2009). Genome-wide analysis in vivo of translation with nucleotide resolution using ribosome profiling. Science *324*, 218-223.

Inouye, M., and Delihas, N. (1988). Small RNAs in the prokaryotes: a growing list of diverse roles. Cell *53*, 5-7.

Iyer, M.K., Niknafs, Y.S., Malik, R., Singhal, U., Sahu, A., Hosono, Y., Barrette, T.R., Prensner, J.R., Evans, J.R., Zhao, S., *et al.* (2015). The landscape of long noncoding RNAs in the human transcriptome. Nat Genet *47*, 199-208.

Jeck, W.R., Sorrentino, J.A., Wang, K., Slevin, M.K., Burd, C.E., Liu, J., Marzluff, W.F., and Sharpless, N.E. (2013). Circular RNAs are abundant, conserved, and associated with ALU repeats. Rna *19*, 141-157.

Jeon, Y., and Lee, J.T. (2011). YY1 tethers Xist RNA to the inactive X nucleation center. Cell *146*, 119-133.

Ji, P., Diederichs, S., Wang, W., Boing, S., Metzger, R., Schneider, P.M., Tidow, N., Brandt, B., Buerger, H., Bulk, E., *et al.* (2003). MALAT-1, a novel noncoding RNA, and thymosin beta4 predict metastasis and survival in early-stage non-small cell lung cancer. Oncogene *22*, 8031-8041.

Jiang, C., Li, Y., Zhao, Z., Lu, J., Chen, H., Ding, N., Wang, G., Xu, J., and Li, X. (2016). Identifying and functionally characterizing tissue-specific and ubiquitously expressed human lncRNAs. Oncotarget *7*, 7120-7133.

Johnsson, P., Lipovich, L., Grander, D., and Morris, K.V. (2014). Evolutionary conservation of long non-coding RNAs; sequence, structure, function. Biochimica et biophysica acta *1840*, 1063-1071.

Kallen, A.N., Zhou, X.B., Xu, J., Qiao, C., Ma, J., Yan, L., Lu, L., Liu, C., Yi, J.S., Zhang, H., *et al.* (2013). The imprinted H19 lncRNA antagonizes let-7 microRNAs. Mol Cell *52*, 101-112.

Kambara, H., Gunawardane, L., Zebrowski, E., Kostadinova, L., Jobava, R., Krokowski, D., Hatzoglou, M., Anthony, D.D., and Valadkhan, S. (2014). Regulation of Interferon-Stimulated Gene BST2 by a lncRNA Transcribed from a Shared Bidirectional Promoter. Frontiers in immunology *5*, 676.

Kapranov, P., Cawley, S.E., Drenkow, J., Bekiranov, S., Strausberg, R.L., Fodor, S.P., and Gingeras, T.R. (2002). Large-scale transcriptional activity in chromosomes 21 and 22. Science *296*, 916-919.

Kapranov, P., Drenkow, J., Cheng, J., Long, J., Helt, G., Dike, S., and Gingeras, T.R. (2005). Examples of the complex architecture of the human transcriptome revealed by RACE and high-density tiling arrays. Genome Res *15*, 987-997.

Kapranov, P., Willingham, A.T., and Gingeras, T.R. (2007). Genome-wide transcription and the implications for genomic organization. Nat Rev Genet *8*, 413-423.

Kapusta, A., Kronenberg, Z., Lynch, V.J., Zhuo, X., Ramsay, L., Bourque, G., Yandell, M., and Feschotte, C. (2013). Transposable elements are major contributors to the origin, diversification, and regulation of vertebrate long noncoding RNAs. PLoS Genet *9*, e1003470.

Katayama, S., Tomaru, Y., Kasukawa, T., Waki, K., Nakanishi, M., Nakamura, M., Nishida, H., Yap, C.C., Suzuki, M., Kawai, J., *et al.* (2005). Antisense transcription in the mammalian transcriptome. Science *309*, 1564-1566.

Khalil, A.M., Guttman, M., Huarte, M., Garber, M., Raj, A., Rivea Morales, D., Thomas, K., Presser, A., Bernstein, B.E., van Oudenaarden, A., *et al.* (2009). Many human large intergenic noncoding RNAs associate with chromatin-modifying complexes and affect gene expression. Proc Natl Acad Sci U S A *106*, 11667-11672.

Kim, T., Cui, R., Jeon, Y.J., Fadda, P., Alder, H., and Croce, C.M. (2015). MYC-repressed long noncoding RNAs antagonize MYC-induced cell proliferation and cell cycle progression. Oncotarget *6*, 18780-18789.

Kim, T., Cui, R., Jeon, Y.J., Lee, J.H., Sim, H., Park, J.K., Fadda, P., Tili, E., Nakanishi, H., Huh, M.I., *et al.* (2014). Long-range interaction and correlation between MYC enhancer and oncogenic long noncoding RNA CARLo-5. Proc Natl Acad Sci U S A *111*, 4173-4178.

Kim, T., Hemberg, M., Gray, J.M., Costa, A.M., Bear, D.M., Wu, J., Harmin, D.A., Laptewicz, M., Barbara-Haley, K., Kuersten, S., *et al.* (2011). Widespread transcription at neuronal activity-regulated enhancers. Nature *465*, 182-187.

Kino, T., Hurt, D.E., Ichijo, T., Nader, N., and Chrousos, G.P. (2010). Noncoding RNA gas5 is a growth arrest- and starvation-associated repressor of the glucocorticoid receptor. Science signaling *3*, ra8.

Kiryu, H., Kin, T., and Asai, K. (2008). Rfold: an exact algorithm for computing local base pairing probabilities. Bioinformatics *24*, 367-373.

Klattenhoff, C.A., Scheuermann, J.C., Surface, L.E., Bradley, R.K., Fields, P.A., Steinhauser, M.L., Ding, H., Butty, V.L., Torrey, L., Haas, S., *et al.* (2013). Braveheart, a long noncoding RNA required for cardiovascular lineage commitment. Cell *152*, 570-583.

Kogo, R., Shimamura, T., Mimori, K., Kawahara, K., Imoto, S., Sudo, T., Tanaka, F., Shibata, K., Suzuki, A., Komune, S., *et al.* (2011). Long noncoding RNA HOTAIR regulates polycomb-dependent chromatin modification and is associated with poor prognosis in colorectal cancers. Cancer Res *71*, 6320-6326.

Kramer, M.C., Liang, D., Tatomer, D.C., Gold, B., March, Z.M., Cherry, S., and Wilusz, J.E. (2015). Combinatorial control of Drosophila circular RNA expression by intronic repeats, hnRNPs, and SR proteins. Genes Dev *29*, 2168-2182.

Kretz, M., Siprashvili, Z., Chu, C., Webster, D.E., Zehnder, A., Qu, K., Lee, C.S., Flockhart, R.J., Groff, A.F., Chow, J., *et al.* (2013). Control of somatic tissue differentiation by the long non-coding RNA TINCR. Nature *493*, 231-235.

Kruger, K., Grabowski, P.J., Zaug, A.J., Sands, J., Gottschling, D.E., and Cech, T.R. (1982). Self-splicing RNA: autoexcision and autocyclization of the ribosomal RNA intervening sequence of Tetrahymena. Cell *31*, 147-157.

Kumar, V., Westra, H.J., Karjalainen, J., Zhernakova, D.V., Esko, T., Hrdlickova, B., Almeida, R., Zhernakova, A., Reinmaa, E., Vosa, U., *et al.* (2013). Human disease-associated genetic variation impacts large intergenic non-coding RNA expression. PLoS Genet *9*, e1003201.

Kundaje, A., Meuleman, W., Ernst, J., Bilenky, M., Yen, A., Heravi-Moussavi, A., Kheradpour, P., Zhang, Z., Wang, J., Ziller, M.J., *et al.* (2015). Integrative analysis of 111 reference human epigenomes. Nature *518*, 317-330.

Kwapisz M, R.M., van Dijk E, Gourvennec S, Descrimes M, Taddei A and Morillon A (2015). Expression of Subtelomeric lncRNAs Links Telomeres Dynamics to RNA Decay in S. cerevisiae. Non-Coding RNA *1*, 94-126.

Lai, F., Gardini, A., Zhang, A., and Shiekhattar, R. (2015). Integrator mediates the biogenesis of enhancer RNAs. Nature *525*, 399-403.

Lai, F., Orom, U.A., Cesaroni, M., Beringer, M., Taatjes, D.J., Blobel, G.A., and Shiekhattar, R. (2013). Activating RNAs associate with Mediator to enhance chromatin architecture and transcription. Nature *494*, 497-501.

Lander, E.S., Linton, L.M., Birren, B., Nusbaum, C., and al, e. (2001a). Initial sequencing and analysis of the human genome. Nature *409*, 860-921.

Lander, E.S., Linton, L.M., Birren, B., Nusbaum, C., Zody, M.C., Baldwin, J., Devon, K., Dewar, K., Doyle, M., FitzHugh, W., *et al.* (2001b). Initial sequencing and analysis of the human genome. Nature *409*, 860-921.

Landerer, E., Villegas, J., Burzio, V.A., Oliveira, L., Villota, C., Lopez, C., Restovic, F., Martinez, R., Castillo, O., and Burzio, L.O. (2011). Nuclear localization of the mitochondrial ncRNAs in normal and cancer cells. Cell Oncol (Dordr) *34*, 297-305.

Lardenois, A., Liu, Y., Walther, T., Chalmel, F., Evrard, B., Granovskaia, M., Chu, A., Davis, R.W., Steinmetz, L.M., and Primig, M. (2010). Execution of the meiotic noncoding RNA expression program and the onset of gametogenesis in yeast require the conserved exosome subunit Rrp6. Proc Natl Acad Sci U S A *108*, 1058-1063.

Latos, P.A., Pauler, F.M., Koerner, M.V., Senergin, H.B., Hudson, Q.J., Stocsits, R.R., Allhoff, W., Stricker, S.H., Klement, R.M., Warczok, K.E., *et al.* (2012). Airn transcriptional overlap, but not its lncRNA products, induces imprinted Igf2r silencing. Science *338*, 1469-1472.

Lazorthes, S., Vallot, C., Briois, S., Aguirrebengoa, M., Thuret, J.Y., St Laurent, G., Rougeulle, C., Kapranov, P., Mann, C., Trouche, D., *et al.* (2015). A vlincRNA participates in senescence maintenance by relieving H2AZ-mediated repression at the INK4 locus. Nat Commun *6*, 5971.

Lee, J.T. (2000). Disruption of imprinted X inactivation by parent-of-origin effects at Tsix. Cell *103*, 17-27.

Lee, J.T., and Bartolomei, M.S. (2013). X-inactivation, imprinting, and long noncoding RNAs in health and disease. Cell *152*, 1308-1323.

Lee, J.T., and Lu, N. (1999). Targeted mutagenesis of Tsix leads to nonrandom X inactivation. Cell *99*, 47-57.

Lee, R.C., Feinbaum, R.L., and Ambros, V. (1993). The C. elegans heterochronic gene lin-4 encodes small RNAs with antisense complementarity to lin-14. Cell *75*, 843-854.

Lee, S., Kopp, F., Chang, T.C., Sataluri, A., Chen, B., Sivakumar, S., Yu, H., Xie, Y., and Mendell, J.T. (2016). Noncoding RNA NORAD Regulates Genomic Stability by Sequestering PUMILIO Proteins. Cell *164*, 69-80.

Lepoivre, C., Belhocine, M., Bergon, A., Griffon, A., Yammine, M., Vanhille, L., Zacarias-Cabeza, J., Garibal, M.A., Koch, F., Maqbool, M.A., *et al.* (2013). Divergent transcription is associated with promoters of transcriptional regulators. BMC genomics *14*, 914.

Leveille, N., Melo, C.A., Rooijers, K., Diaz-Lagares, A., Melo, S.A., Korkmaz, G., Lopes, R., Akbari Moqadam, F., Maia, A.R., Wijchers, P.J., *et al.* (2015). Genome-wide profiling of p53-regulated enhancer RNAs uncovers a subset of enhancers controlled by a lncRNA. Nat Commun *6*, 6520.

Lewis, J.B., Atkins, J.F., Anderson, C.W., Baum, P.R., and Gesteland, R.F. (1975). Mapping of late adenovirus genes by cell-free translation of RNA selected by hybridization to specific DNA fragments. Proc Natl Acad Sci U S A *72*, 1344-1348.

Li, F., Xiao, Y., Huang, F., Deng, W., Zhao, H., Shi, X., Wang, S., Yu, X., Zhang, L., Han, Z., *et al.* (2015a). Spatiotemporal-specific lncRNAs in the brain, colon, liver and lung of macaque during development. Molecular bioSystems *11*, 3253-3263.

Li, J., Poursat, M.A., Drubay, D., Motz, A., Saci, Z., Morillon, A., Michiels, S., and Gautheret, D. (2015b). A Dual Model for Prioritizing Cancer Mutations in the Non-coding Genome Based on Germline and Somatic Events. PLoS computational biology *11*, e1004583.

Li, J., Yang, J., Zhou, P., Le, Y., Zhou, C., Wang, S., Xu, D., Lin, H.K., and Gong, Z. (2015c). Circular RNAs in cancer: novel insights into origins, properties, functions and implications. American journal of cancer research *5*, 472-480.

Li, L., and Chang, H.Y. (2014). Physiological roles of long noncoding RNAs: insight from knockout mice. Trends in cell biology *24*, 594-602.

Li, L., Liu, B., Wapinski, O.L., Tsai, M.C., Qu, K., Zhang, J., Carlson, J.C., Lin, M., Fang, F., Gupta, R.A., *et al.* (2013a). Targeted disruption of Hotair leads to homeotic transformation and gene derepression. Cell reports *5*, 3-12.

Li, M., Gou, H., Tripathi, B.K., Huang, J., Jiang, S., Dubois, W., Waybright, T., Lei, M., Shi, J., and Zhou, M. (2015d). An Apela RNA-Containing Negative Feedback Loop Regulates p53-Mediated Apoptosis in Embryonic Stem Cells. Cell stem cell *16*, 669-683.

Li, W., Notani, D., Ma, Q., Tanasa, B., Nunez, E., Chen, A.Y., Merkurjev, D., Zhang, J., Ohgi, K., Song, X., *et al.* (2013b). Functional roles of enhancer RNAs for oestrogen-dependent transcriptional activation. Nature *498*, 516-520.

Li, W., Notani, D., and Rosenfeld, M.G. (2016). Enhancers as non-coding RNA transcription units: recent insights and future perspectives. Nat Rev Genet *17*, 207-223.

Li, Z., Huang, C., Bao, C., Chen, L., Lin, M., Wang, X., Zhong, G., Yu, B., Hu, W., Dai, L., *et al.* (2015e). Exon-intron circular RNAs regulate transcription in the nucleus. Nat Struct Mol Biol *22*, 256-264.

Lin, R., Maeda, S., Liu, C., Karin, M., and Edgington, T.S. (2007). A large noncoding RNA is a marker for murine hepatocellular carcinomas and a spectrum of human carcinomas. Oncogene *26*, 851-858.

Lindblad-Toh, K., Garber, M., Zuk, O., Lin, M.F., Parker, B.J., Washietl, S., Kheradpour, P., Ernst, J., Jordan, G., Mauceli, E., *et al.* (2011). A high-resolution map of human evolutionary constraint using 29 mammals. Nature *478*, 476-482.

Ling, H., Vincent, K., Pichler, M., Fodde, R., Berindan-Neagoe, I., Slack, F.J., and Calin, G.A. (2015). Junk DNA and the long non-coding RNA twist in cancer genetics. Oncogene *34*, 5003-5011.

Liu, B., Sun, L., Liu, Q., Gong, C., Yao, Y., Lv, X., Lin, L., Yao, H., Su, F., Li, D., *et al.* (2015). A cytoplasmic NF-kappaB interacting long noncoding RNA blocks IkappaB phosphorylation and suppresses breast cancer metastasis. Cancer cell *27*, 370-381.

Liu, X.H., Liu, Z.L., Sun, M., Liu, J., Wang, Z.X., and De, W. (2013). The long non-coding RNA HOTAIR indicates a poor prognosis and promotes metastasis in non-small cell lung cancer. BMC cancer *13*, 464.

Liz, J., Portela, A., Soler, M., Gomez, A., Ling, H., Michlewski, G., Calin, G.A., Guil, S., and Esteller, M. (2014). Regulation of pri-miRNA processing by a long noncoding RNA transcribed from an ultraconserved region. Mol Cell *55*, 138-147.

Loewer, S., Cabili, M.N., Guttman, M., Loh, Y.H., Thomas, K., Park, I.H., Garber, M., Curran, M., Onder, T., Agarwal, S., *et al.* (2010). Large intergenic non-coding RNA-RoR modulates reprogramming of human induced pluripotent stem cells. Nat Genet *42*, 1113-1117.

Lu, Z., and Chang, H.Y. (2016). Decoding the RNA structurome. Current opinion in structural biology *36*, 142-148.

Lu, Z., Xiao, Z., Liu, F., Cui, M., Li, W., Yang, Z., Li, J., Ye, L., and Zhang, X. (2016). Long non-coding RNA HULC promotes tumor angiogenesis in liver cancer by up-regulating sphingosine kinase 1 (SPHK1). Oncotarget *7*, 241-254.

Lyon, M.F. (1961). Gene action in the X-chromosome of the mouse (Mus musculus L.). Nature *190*, 372-373.

Maamar, H., Cabili, M.N., Rinn, J., and Raj, A. (2013). linc-HOXA1 is a noncoding RNA that represses Hoxa1 transcription in cis. Genes Dev *27*, 1260-1271.

Magistri, M., Faghihi, M.A., St Laurent, G., 3rd, and Wahlestedt, C. (2012). Regulation of chromatin structure by long noncoding RNAs: focus on natural antisense transcripts. Trends in genetics : TIG *28*, 389-396.

Mancini-Dinardo, D., Steele, S.J., Levorse, J.M., Ingram, R.S., and Tilghman, S.M. (2006). Elongation of the Kcnq1ot1 transcript is required for genomic imprinting of neighboring genes. Genes Dev *20*, 1268-1282.

Marin-Bejar, O., Marchese, F.P., Athie, A., Sanchez, Y., Gonzalez, J., Segura, V., Huang, L., Moreno, I., Navarro, A., Monzo, M., *et al.* (2013). Pint lincRNA connects the p53 pathway with epigenetic silencing by the Polycomb repressive complex 2. Genome Biol *14*, R104.

Marques, A.C., Hughes, J., Graham, B., Kowalczyk, M.S., Higgs, D.R., and Ponting, C.P. (2013). Chromatin signatures at transcriptional start sites separate two equally populated yet distinct classes of intergenic long noncoding RNAs. Genome Biol *14*, R131.

Marques, A.C., and Ponting, C.P. (2014). Intergenic lncRNAs and the evolution of gene expression. Current opinion in genetics & development *27*, 48-53.

Massone, S., Ciarlo, E., Vella, S., Nizzari, M., Florio, T., Russo, C., Cancedda, R., and Pagano, A. (2012). NDM29, a RNA polymerase III-dependent non coding RNA, promotes amyloidogenic processing of APP and amyloid beta secretion. Biochimica et biophysica acta *1823*, 1170-1177.

Mattick, J.S. (1994). Introns: evolution and function. Curr Opin Genet Dev. Dec; 4(6):823-31.

Mattick, J.S. (2001). Non-coding RNAs: the architects of eukaryotic complexity. EMBO Rep *2*, 986-991.

Mattick, J.S. (2003). Challenging the dogma: the hidden layer of non-protein-coding RNAs in complex organisms. BioEssays : news and reviews in molecular, cellular and developmental biology *25*, 930-939.

Mattick, J.S. (2004). RNA regulation: a new genetics? Nat Rev Genet. Apr; 5(4):316-23

Mattick, J.S., and Gagen, M.J. (2001). The evolution of controlled multitasked gene networks: the role of introns and other noncoding RNAs in the development of complex organisms. Molecular biology and evolution *18*, 1611-1630.

Mattick, J.S., and Rinn, J.L. (2015). Discovery and annotation of long noncoding RNAs. Nat Struct Mol Biol *22*, 5-7.

Maurano, M.T., Humbert, R., Rynes, E., Thurman, R.E., Haugen, E., Wang, H., Reynolds, A.P., Sandstrom, R., Qu, H., Brody, J., *et al.* (2012). Systematic localization of common disease-associated variation in regulatory DNA. Science *337*, 1190-1195.

Melton, C., Reuter, J.A., Spacek, D.V., and Snyder, M. (2015). Recurrent somatic mutations in regulatory regions of human cancer genomes. Nat Genet *47*, 710-716.

Memczak, S., Jens, M., Elefsinioti, A., Torti, F., Krueger, J., Rybak, A., Maier, L., Mackowiak, S.D., Gregersen, L.H., Munschauer, M., *et al.* (2013). Circular RNAs are a large class of animal RNAs with regulatory potency. Nature *495*, 333-338.

Mercer, T.R., Wilhelm, D., Dinger, M.E., Solda, G., Korbie, D.J., Glazov, E.A., Truong, V., Schwenke, M., Simons, C., Matthaei, K.I., *et al.* (2011). Expression of distinct RNAs from 3' untranslated regions. Nucleic Acids Res *39*, 2393-2403.

Meseure, D., Vacher, S., Lallemand, F., Alsibai, K.D., Hatem, R., Chemlali, W., Nicolas, A., De Koning, L., Pasmant, E., Callens, C., *et al.* (2016). Prognostic value of a newly identified MALAT1 alternatively spliced transcript in breast cancer. British journal of cancer *114*, 1395-1404.

Mestdagh, P., Fredlund, E., Pattyn, F., Rihani, A., Van Maerken, T., Vermeulen, J., Kumps, C., Menten, B., De Preter, K., Schramm, A., *et al.* (2010). An integrative genomics screen uncovers ncRNA T-UCR functions in neuroblastoma tumours. Oncogene *29*, 3583-3592.

Michalik, K.M., You, X., Manavski, Y., Doddaballapur, A., Zornig, M., Braun, T., John, D., Ponomareva, Y., Chen, W., Uchida, S., *et al.* (2014). Long noncoding RNA MALAT1 regulates endothelial cell function and vessel growth. Circulation research *114*, 1389-1397.

Migeon, B.R., Lee, C.H., Chowdhury, A.K., and Carpenter, H. (2002). Species differences in TSIX/Tsix reveal the roles of these genes in X-chromosome inactivation. American journal of human genetics *71*, 286-293.

Milligan, M.J., and Lipovich, L. (2014). Pseudogene-derived lncRNAs: emerging regulators of gene expression. Frontiers in genetics *5*, 476.

Modarresi, F., Faghihi, M.A., Lopez-Toledano, M.A., Fatemi, R.P., Magistri, M., Brothers, S.P., van der Brug, M.P., and Wahlestedt, C. (2012). Inhibition of natural antisense transcripts in vivo results in gene-specific transcriptional upregulation. Nat Biotechnol *30*, 453-459.

Mondal, T., Rasmussen, M., Pandey, G.K., Isaksson, A., and Kanduri, C. (2010). Characterization of the RNA content of chromatin. Genome Res *20*, 899-907.

Mondal, T., Subhash, S., Vaid, R., Enroth, S., Uday, S., Reinius, B., Mitra, S., Mohammed, A., James, A.R., Hoberg, E., *et al.* (2015). MEG3 long noncoding RNA regulates the TGF-beta pathway genes through formation of RNA-DNA triplex structures. Nat Commun *6*, 7743.

Montes, M., Nielsen, M.M., Maglieri, G., Jacobsen, A., Hojfeldt, J., Agrawal-Singh, S., Hansen, K., Helin, K., van de Werken, H.J., Pedersen, J.S., *et al.* (2015). The lncRNA MIR31HG regulates p16(INK4A) expression to modulate senescence. Nat Commun *6*, 6967.

Montgomery, M.K. (2004). RNA interference: historical overview and significance. Methods Mol Biol *265*, 3-21.

Morris, K.V., and Mattick, J.S. (2014). The rise of regulatory RNA. Nat Rev Genet *15*, 423-437.

Morris, K.V., Santoso, S., Turner, A.M., Pastori, C., and Hawkins, P.G. (2008). Bidirectional transcription directs both transcriptional gene activation and suppression in human cells. PLoS Genet *4*, e1000258.

Mourtada-Maarabouni, M., Pickard, M.R., Hedge, V.L., Farzaneh, F., and Williams, G.T. (2009). GAS5, a non-protein-coding RNA, controls apoptosis and is downregulated in breast cancer. Oncogene *28*, 195-208.

Murray, S.C., Haenni, S., Howe, F.S., Fischl, H., Chocian, K., Nair, A., and Mellor, J. (2015). Sense and antisense transcription are associated with distinct chromatin architectures across genes. Nucleic Acids Res *43*, 7823-7837.

Nakagawa, S., Ip, J.Y., Shioi, G., Tripathi, V., Zong, X., Hirose, T., and Prasanth, K.V. (2012). Malat1 is not an essential component of nuclear speckles in mice. Rna *18*, 1487-1499.

Nakaya, H.I., Amaral, P.P., Louro, R., Lopes, A., Fachel, A.A., Moreira, Y.B., El-Jundi, T.A., da Silva, A.M., Reis, E.M., and Verjovski-Almeida, S. (2007). Genome mapping and expression analyses of human intronic noncoding RNAs reveal tissue-specific patterns and enrichment in genes related to regulation of transcription. Genome Biol *8*, R43.

Nam, J.W., and Bartel, D.P. (2012). Long noncoding RNAs in C. elegans. Genome Res *22*, 2529-2540.

Ne, I.I., Heward, J.A., Roux, B., Tsitsiou, E., Fenwick, P.S., Lenzi, L., Goodhead, I., Hertz-Fowler, C., Heger, A., Hall, N., *et al.* (2014). Long non-coding RNAs and enhancer RNAs regulate the lipopolysaccharide-induced inflammatory response in human monocytes. Nat Commun *5*, 3979.

Necsulea, A., Soumillon, M., Warnefors, M., Liechti, A., Daish, T., Zeller, U., Baker, J.C., Grutzner, F., and Kaessmann, H. (2014). The evolution of lncRNA repertoires and expression patterns in tetrapods. Nature *505*, 635-640.

Neil, H., Malabat, C., d'Aubenton-Carafa, Y., Xu, Z., Steinmetz, L.M., and Jacquier, A. (2009). Widespread bidirectional promoters are the major source of cryptic transcripts in yeast. Nature *457*, 1038-1042.

Ng, S.Y., Johnson, R., and Stanton, L.W. (2012). Human long non-coding RNAs promote pluripotency and neuronal differentiation by association with chromatin modifiers and transcription factors. Embo J *31*, 522-533.

Noh, J.H., Kim, K.M., Abdelmohsen, K., Yoon, J.H., Panda, A.C., Munk, R., Kim, J., Curtis, J., Moad, C.A., Wohler, C.M., *et al.* (2016). HuR and GRSF1 modulate the nuclear export and mitochondrial localization of the lncRNA RMRP. Genes Dev *30*, 1224-1239.

Nozawa, R.S., Nagao, K., Igami, K.T., Shibata, S., Shirai, N., Nozaki, N., Sado, T., Kimura, H., and Obuse, C. (2013). Human inactive X chromosome is compacted through a PRC2-independent SMCHD1-HBiX1 pathway. Nat Struct Mol Biol *20*, 566-573.

Ntini, E., Jarvelin, A.I., Bornholdt, J., Chen, Y., Boyd, M., Jorgensen, M., Andersson, R., Hoof, I., Schein, A., Andersen, P.R., *et al.* (2013). Polyadenylation site-induced decay of upstream transcripts enforces promoter directionality. Nat Struct Mol Biol *20*, 923-928.

Okazaki, Y., Furuno, M., Kasukawa, T., Adachi, J., Bono, H., Kondo, S., Nikaido, I., Osato, N., Saito, R., Suzuki, H., *et al.* (2002). Analysis of the mouse transcriptome based on functional annotation of 60,770 full-length cDNAs. Nature *420*, 563-573.

Orgel, L.E., and Crick, F.H. (1980). Selfish DNA: the ultimate parasite. Nature *284*, 604-607.

Orom, U.A., Derrien, T., Beringer, M., Gumireddy, K., Gardini, A., Bussotti, G., Lai, F., Zytnicki, M., Notredame, C., Huang, Q., *et al.* (2010). Long noncoding RNAs with enhancer-like function in human cells. Cell *143*, 46-58.

Pachnis, V., Belayew, A., and Tilghman, S.M. (1984). Locus unlinked to alpha-fetoprotein under the control of the murine raf and Rif genes. Proc Natl Acad Sci U S A *81*, 5523-5527.

Pandey, G.K., Mitra, S., Subhash, S., Hertwig, F., Kanduri, M., Mishra, K., Fransson, S., Ganeshram, A., Mondal, T., Bandaru, S., *et al.* (2014). The risk-associated long noncoding RNA NBAT-1 controls neuroblastoma progression by regulating cell proliferation and neuronal differentiation. Cancer cell *26*, 722-737.

Pek, J.W., Osman, I., Tay, M.L., and Zheng, R.T. (2015). Stable intronic sequence RNAs have possible regulatory roles in Drosophila melanogaster. J Cell Biol *211*, 243-251.

Peng, L., Yuan, X.Q., and Li, G.C. (2015). The emerging landscape of circular RNA ciRS-7 in cancer (Review). Oncology reports *33*, 2669-2674.

Penny, G.D., Kay, G.F., Sheardown, S.A., Rastan, S., and Brockdorff, N. (1996). Requirement for Xist in X chromosome inactivation. Nature *379*, 131-137.

Poliseno, L., Salmena, L., Zhang, J., Carver, B., Haveman, W.J., and Pandolfi, P.P. (2010). A coding-independent function of gene and pseudogene mRNAs regulates tumour biology. Nature *465*, 1033-1038.

Pollard, K.S., Salama, S.R., King, B., Kern, A.D., Dreszer, T., Katzman, S., Siepel, A., Pedersen, J.S., Bejerano, G., Baertsch, R., *et al.* (2006a). Forces shaping the fastest evolving regions in the human genome. PLoS Genet *2*, e168.

Pollard, K.S., Salama, S.R., Lambert, N., Lambot, M.A., Coppens, S., Pedersen, J.S., Katzman, S., King, B., Onodera, C., Siepel, A., *et al.* (2006b). An RNA gene expressed during cortical development evolved rapidly in humans. Nature *443*, 167-172.

Ponting, C., Oliver, P., and Reik, W. (2009). Evolution and functions of long noncoding RNAs. Cell *136*, 629-641.

Porro, A., Feuerhahn, S., Reichenbach, P., and Lingner, J. (2010). Molecular dissection of telomeric repeat-containing RNA biogenesis unveils the presence of distinct and multiple regulatory pathways. Mol Cell Biol *30*, 4808-4817.

Porrua, O., and Libri, D. (2015). Transcription termination and the control of the transcriptome: why, where and how to stop. Nature reviews Molecular cell biology *16*, 190-202.

Postepska-Igielska, A., Giwojna, A., Gasri-Plotnitsky, L., Schmitt, N., Dold, A., Ginsberg, D., and Grummt, I. (2015). LncRNA Khps1 Regulates Expression of the Proto-oncogene SPHK1 via Triplex-Mediated Changes in Chromatin Structure. Mol Cell *60*, 626-636.

Preker, P., Almvig, K., Christensen, M.S., Valen, E., Mapendano, C.K., Sandelin, A., and Jensen, T.H. (2011). PROMoter uPstream Transcripts share characteristics with mRNAs and are produced upstream of all three major types of mammalian promoters. Nucleic Acids Res *39*, 7179-7193.

Preker, P., Nielsen, J., Kammler, S., Lykke-Andersen, S., Christensen, M.S., Mapendano, C.K., Schierup, M.H., and Jensen, T.H. (2008). RNA exosome depletion reveals transcription upstream of active human promoters. Science *322*, 1851-1854.

Prensner, J.R., Iyer, M.K., Balbin, O.A., Dhanasekaran, S.M., Cao, Q., Brenner, J.C., Laxman, B., Asangani, I.A., Grasso, C.S., Kominsky, H.D., *et al.* (2011). Transcriptome sequencing across a prostate cancer cohort identifies PCAT-1, an unannotated lincRNA implicated in disease progression. Nat Biotechnol *29*, 742-749.

Prensner, J.R., Iyer, M.K., Sahu, A., Asangani, I.A., Cao, Q., Patel, L., Vergara, I.A., Davicioni, E., Erho, N., Ghadessi, M., *et al.* (2013). The long noncoding RNA SChLAP1 promotes aggressive prostate cancer and antagonizes the SWI/SNF complex. Nat Genet *45*, 1392-1398.

Prensner, J.R., Zhao, S., Erho, N., Schipper, M., Iyer, M.K., Dhanasekaran, S.M., Magi-Galluzzi, C., Mehra, R., Sahu, A., Siddiqui, J., *et al.* (2014). RNA biomarkers associated with metastatic progression in prostate cancer: a multi-institutional high-throughput analysis of SChLAP1. The Lancet Oncology *15*, 1469-1480.

Puvvula, P.K., Desetty, R.D., Pineau, P., Marchio, A., Moon, A., Dejean, A., and Bischof, O. (2014). Long noncoding RNA PANDA and scaffold-attachment-factor SAFA control senescence entry and exit. Nat Commun *5*, 5323.

Quenet, D., and Dalal, Y. (2014). A long non-coding RNA is required for targeting centromeric protein A to the human centromere. eLife *3*, e03254.

Quinn, J.J., and Chang, H.Y. (2016). Unique features of long non-coding RNA biogenesis and function. Nat Rev Genet *17*, 47-62.

Qureshi, I.A., and Mehler, M.F. (2012). Emerging roles of non-coding RNAs in brain evolution, development, plasticity and disease. Nature reviews Neuroscience *13*, 528-541.

Rackham, O., Shearwood, A.M., Mercer, T.R., Davies, S.M., Mattick, J.S., and Filipovska, A. (2011). Long noncoding RNAs are generated from the mitochondrial genome and regulated by nuclear-encoded proteins. Rna *17*, 2085-2093.

Rands, C.M., Meader, S., Ponting, C.P., and Lunter, G. (2014). 8.2% of the Human genome is constrained: variation in rates of turnover across functional element classes in the human lineage. PLoS Genet *10*, e1004525.

Rapicavoli, N.A., Qu, K., Zhang, J., Mikhail, M., Laberge, R.M., and Chang, H.Y. (2013). A mammalian pseudogene lncRNA at the interface of inflammation and anti-inflammatory therapeutics. eLife *2*, e00762.

Rinn, J.L., Kertesz, M., Wang, J.K., Squazzo, S.L., Xu, X., Brugmann, S.A., Goodnough, L.H., Helms, J.A., Farnham, P.J., Segal, E., *et al.* (2007). Functional Demarcation of Active and Silent Chromatin Domains in Human HOX Loci by Noncoding RNAs. Cell *129*, 1311-1323.

Rippe, K., and Luke, B. (2015). TERRA and the state of the telomere. Nat Struct Mol Biol *22*, 853-858.

Roest Crollius, H., Jaillon, O., Dasilva, C., Ozouf-Costaz, C., Fizames, C., Fischer, C., Bouneau, L., Billault, A., Quetier, F., Saurin, W., *et al.* (2000). Characterization and repeat analysis of the compact genome of the freshwater pufferfish Tetraodon nigroviridis. Genome Res *10*, 939-949.

Rosic, S., Kohler, F., and Erhardt, S. (2014). Repetitive centromeric satellite RNA is essential for kinetochore formation and cell division. J Cell Biol *207*, 335-349.

Rybak-Wolf, A., Stottmeister, C., Glazar, P., Jens, M., Pino, N., Giusti, S., Hanan, M., Behm, M., Bartok, O., Ashwal-Fluss, R., *et al.* (2015). Circular RNAs in the Mammalian Brain Are Highly Abundant, Conserved, and Dynamically Expressed. Mol Cell *58*, 870-885.

Sado, T., Wang, Z., Sasaki, H., and Li, E. (2001). Regulation of imprinted X-chromosome inactivation in mice by Tsix. Development *128*, 1275-1286.

Salzman, J., Gawad, C., Wang, P.L., Lacayo, N., and Brown, P.O. (2012). Circular RNAs are the predominant transcript isoform from hundreds of human genes in diverse cell types. PLoS One *7*, e30733.

Sanchez, Y., Segura, V., Marin-Bejar, O., Athie, A., Marchese, F.P., Gonzalez, J., Bujanda, L., Guo, S., Matheu, A., and Huarte, M. (2014). Genome-wide analysis of the human p53 transcriptional network unveils a lncRNA tumour suppressor signature. Nat Commun *5*, 5812.

Sanger, F., Coulson, A.R., Friedmann, T., Air, G.M., Barrell, B.G., Brown, N.L., Fiddes, J.C., Hutchison, C.A., 3rd, Slocombe, P.M., and Smith, M. (1978). The nucleotide sequence of bacteriophage phiX174. Journal of molecular biology *125*, 225-246.

Satpathy, A.T., and Chang, H.Y. (2015). Long noncoding RNA in hematopoiesis and immunity. Immunity *42*, 792-804.

Sauvageau, M., Goff, L.A., Lodato, S., Bonev, B., Groff, A.F., Gerhardinger, C., Sanchez-Gomez, D.B., Hacisuleyman, E., Li, E., Spence, M., *et al.* (2013). Multiple knockout mouse models reveal lincRNAs are required for life and brain development. eLife *2*, e01749.

Scaruffi, P., Stigliani, S., Moretti, S., Coco, S., De Vecchi, C., Valdora, F., Garaventa, A., Bonassi, S., and Tonini, G.P. (2009). Transcribed-Ultra Conserved Region expression is associated with outcome in high-risk neuroblastoma. BMC cancer *9*, 441.

Schoeftner, S., Sengupta, A.K., Kubicek, S., Mechtler, K., Spahn, L., Koseki, H., Jenuwein, T., and Wutz, A. (2006). Recruitment of PRC1 function at the initiation of X inactivation independent of PRC2 and silencing. Embo J *25*, 3110-3122.

Schorderet, P., and Duboule, D. (2011). Structural and functional differences in the long non-coding RNA hotair in mouse and human. PLoS Genet *7*, e1002071.

Schulz, D., Schwalb, B., Kiesel, A., Baejen, C., Torkler, P., Gagneur, J., Soeding, J., and Cramer, P. (2013). Transcriptome surveillance by selective termination of noncoding RNA synthesis. Cell *155*, 1075-1087.

Scruggs, B.S., Gilchrist, D.A., Nechaev, S., Muse, G.W., Burkholder, A., Fargo, D.C., and Adelman, K. (2015). Bidirectional Transcription Arises from Two Distinct Hubs of Transcription Factor Binding and Active Chromatin. Mol Cell *58*, 1101-1112.

Seila, A.C., Calabrese, J.M., Levine, S.S., Yeo, G.W., Rahl, P.B., Flynn, R.A., Young, R.A., and Sharp, P.A. (2008). Divergent transcription from active promoters. Science *322*, 1849-1851.

Shahryari, A., Jazi, M.S., Samaei, N.M., and Mowla, S.J. (2015). Long non-coding RNA SOX2OT: expression signature, splicing patterns, and emerging roles in pluripotency and tumorigenesis. Frontiers in genetics *6*, 196.

Shao, Y., Ye, M., Jiang, X., Sun, W., Ding, X., Liu, Z., Ye, G., Zhang, X., Xiao, B., and Guo, J. (2014). Gastric juice long noncoding RNA used as a tumor marker for screening gastric cancer. Cancer *120*, 3320-3328.

Sheik Mohamed, J., Gaughwin, P.M., Lim, B., Robson, P., and Lipovich, L. (2010). Conserved long noncoding RNAs transcriptionally regulated by Oct4 and Nanog modulate pluripotency in mouse embryonic stem cells. Rna *16*, 324-337.

Sigova, A.A., Mullen, A.C., Molinie, B., Gupta, S., Orlando, D.A., Guenther, M.G., Almada, A.E., Lin, C., Sharp, P.A., Giallourakis, C.C., *et al.* (2013). Divergent transcription of long noncoding RNA/mRNA gene pairs in embryonic stem cells. Proc Natl Acad Sci U S A *110*, 2876-2881.

Singh, D.K., and Prasanth, K.V. (2013). Functional insights into the role of nuclear-retained long noncoding RNAs in gene expression control in mammalian cells. Chromosome research : an international journal on the molecular, supramolecular and evolutionary aspects of chromosome biology *21*, 695-711.

Sleutels, F., Zwart, R., and Barlow, D.P. (2002). The non-coding Air RNA is required for silencing autosomal imprinted genes. Nature *415*, 810-813.

Somarowthu, S., Legiewicz, M., Chillon, I., Marcia, M., Liu, F., and Pyle, A.M. (2015). HOTAIR forms an intricate and modular secondary structure. Mol Cell *58*, 353-361.

Song, X., Wang, X., Arai, S., and Kurokawa, R. (2012). Promoter-associated noncoding RNA from the CCND1 promoter. Methods Mol Biol *809*, 609-622.

Spitale, R.C., Flynn, R.A., Zhang, Q.C., Crisalli, P., Lee, B., Jung, J.W., Kuchelmeister, H.Y., Batista, P.J., Torre, E.A., Kool, E.T., *et al.* (2015). Structural imprints in vivo decode RNA regulatory mechanisms. Nature *519*, 486-490.

Spurlock, C.F., 3rd, Tossberg, J.T., Guo, Y., Collier, S.P., Crooke, P.S., 3rd, and Aune, T.M. (2015). Expression and functions of long noncoding RNAs during human T helper cell differentiation. Nat Commun *6*, 6932.

St Laurent, G., Shtokalo, D., Dong, B., Tackett, M.R., Fan, X., Lazorthes, S., Nicolas, E., Sang, N., Triche, T.J., McCaffrey, T.A., *et al.* (2013). VlincRNAs controlled by retroviral elements are a hallmark of pluripotency and cancer. Genome Biol *14*, R73.

St Laurent, G., Shtokalo, D., Tackett, M.R., Yang, Z., Eremina, T., Wahlestedt, C., Urcuqui-Inchima, S., Seilheimer, B., McCaffrey, T.A., and Kapranov, P. (2012). Intronic RNAs constitute the major fraction of the non-coding RNA in mammalian cells. BMC genomics *13*, 504.

St Laurent, G., Vyatkin, Y., Antonets, D., Ri, M., Qi, Y., Saik, O., Shtokalo, D., de Hoon, M.J., Kawaji, H., Itoh, M., *et al.* (2016). Functional annotation of the vlinc class of non-coding RNAs using systems biology approach. Nucleic Acids Res *44*, 3233-3252.

St Laurent, G., Wahlestedt, C., and Kapranov, P. (2015). The Landscape of long noncoding RNA classification. Trends in genetics : TIG *31*, 239-251.

Stamatoyannopoulos, J.A., Snyder, M., Hardison, R., Ren, B., Gingeras, T., Gilbert, D.M., Groudine, M., Bender, M., Kaul, R., Canfield, T., *et al.* (2012). An encyclopedia of mouse DNA elements (Mouse ENCODE). Genome Biol *13*, 418.

Su, W.Y., Xiong, H., and Fang, J.Y. (2010). Natural antisense transcripts regulate gene expression in an epigenetic manner. Biochemical and biophysical research communications *396*, 177-181.

Suemori, H., and Noguchi, S. (2000). Hox C cluster genes are dispensable for overall body plan of mouse embryonic development. Developmental biology *220*, 333-342.

Sun, L., Goff, L.A., Trapnell, C., Alexander, R., Lo, K.A., Hacisuleyman, E., Sauvageau, M., Tazon-Vega, B., Kelley, D.R., Hendrickson, D.G., *et al.* (2013). Long noncoding RNAs regulate adipogenesis. Proc Natl Acad Sci U S A *110*, 3387-3392.

Szczesniak, M.W., and Makalowska, I. (2016). lncRNA-RNA Interactions across the Human Transcriptome. PLoS One *11*, e0150353.

Takayama, K., Horie-Inoue, K., Katayama, S., Suzuki, T., Tsutsumi, S., Ikeda, K., Urano, T., Fujimura, T., Takagi, K., Takahashi, S., *et al.* (2013). Androgen-responsive long noncoding RNA CTBP1-AS promotes prostate cancer. Embo J *32*, 1665-1680.

Talhouarne, G.J., and Gall, J.G. (2014). Lariat intronic RNAs in the cytoplasm of Xenopus tropicalis oocytes. Rna *20*, 1476-1487.

Tang, L., Zhang, W., Su, B., and Yu, B. (2013). Long noncoding RNA HOTAIR is associated with motility, invasion, and metastatic potential of metastatic melanoma. BioMed research international *2013*, 251098.

Tay, Y., Rinn, J., and Pandolfi, P.P. (2014). The multilayered complexity of ceRNA crosstalk and competition. Nature *505*, 344-352.

Teschendorff, A.E., Lee, S.H., Jones, A., Fiegl, H., Kalwa, M., Wagner, W., Chindera, K., Evans, I., Dubeau, L., Orjalo, A., *et al.* (2015). HOTAIR and its surrogate DNA methylation signature indicate carboplatin resistance in ovarian cancer. Genome medicine *7*, 108.

Trimarchi, T., Bilal, E., Ntziachristos, P., Fabbri, G., Dalla-Favera, R., Tsirigos, A., and Aifantis, I. (2014). Genome-wide mapping and characterization of Notch-regulated long noncoding RNAs in acute leukemia. Cell *158*, 593-606.

Trofimova, I., Chervyakova, D., and Krasikova, A. (2015). Transcription of subtelomere tandemly repetitive DNA in chicken embryogenesis. Chromosome research : an international journal on the molecular, supramolecular and evolutionary aspects of chromosome biology *23*, 495-503.

Tsai, M.C., Manor, O., Wan, Y., Mosammaparast, N., Wang, J.K., Lan, F., Shi, Y., Segal, E., and Chang, H.Y. (2010). Long noncoding RNA as modular scaffold of histone modification complexes. Science *329*, 689-693.

Tseng, Y.Y., Moriarity, B.S., Gong, W., Akiyama, R., Tiwari, A., Kawakami, H., Ronning, P., Reuland, B., Guenther, K., Beadnell, T.C., *et al.* (2014). PVT1 dependence in cancer with MYC copy-number increase. Nature *512*, 82-86.

Twayana, S., Legnini, I., Cesana, M., Cacchiarelli, D., Morlando, M., and Bozzoni, I. (2013). Biogenesis and function of non-coding RNAs in muscle differentiation and in Duchenne muscular dystrophy. Biochemical Society transactions *41*, 844-849.

Uesaka, M., Nishimura, O., Go, Y., Nakashima, K., Agata, K., and Imamura, T. (2014). Bidirectional promoters are the major source of gene activation-associated non-coding RNAs in mammals. BMC genomics *15*, 35.

Ulitsky, I., Shkumatava, A., Jan, C.H., Sive, H., and Bartel, D.P. (2011). Conserved function of lincRNAs in vertebrate embryonic development despite rapid sequence evolution. Cell *147*, 1537-1550.

Ulveling, D., Francastel, C., and Hube, F. (2011). Identification of potentially new bifunctional RNA based on genome-wide data-mining of alternative splicing events. Biochimie *93*, 2024-2027.

Vallot, C., Huret, C., Lesecque, Y., Resch, A., Oudrhiri, N., Bennaceur-Griscelli, A., Duret, L., and Rougeulle, C. (2013). XACT, a long noncoding transcript coating the active X chromosome in human pluripotent cells. Nat Genet *45*, 239-241.

Van Dijk, E.L., Chen, C.L., d'Aubenton-Carafa, Y., Gourvennec, S., Kwapisz, M., Roche, V., Bertrand, C., Silvain, M., Legoix-Né, P., Loeillet, S., *et al.* (2011). XUTs are a class of Xrn1-sensitive antisense regulatory non coding RNA in yeast. Nature *475*, 114-117.

van Heesch, S., van Iterson, M., Jacobi, J., Boymans, S., Essers, P.B., de Bruijn, E., Hao, W., MacInnes, A.W., Cuppen, E., and Simonis, M. (2014). Extensive localization of long noncoding RNAs to the cytosol and mono- and polyribosomal complexes. Genome Biol *15*, R6.

Vembar, S.S., Scherf, A., and Siegel, T.N. (2014). Noncoding RNAs as emerging regulators of Plasmodium falciparum virulence gene expression. Current opinion in microbiology *20*, 153-161.

Venter, J.C., Adams, M.D., Myers, E.W., Li, P.W., Mural, R.J., Sutton, G.G., Smith, H.O., Yandell, M., Evans, C.A., Holt, R.A., *et al.* (2001). The sequence of the human genome. Science *291*, 1304-1351.

Wan, Y., Qu, K., Zhang, Q.C., Flynn, R.A., Manor, O., Ouyang, Z., Zhang, J., Spitale, R.C., Snyder, M.P., Segal, E., *et al.* (2014). Landscape and variation of RNA secondary structure across the human transcriptome. Nature *505*, 706-709.

Wang, D., Garcia-Bassets, I., Benner, C., Li, W., Su, X., Zhou, Y., Qiu, J., Liu, W., Kaikkonen, M.U., Ohgi, K.A., *et al.* (2011a). Reprogramming transcription by distinct classes of enhancers functionally defined by eRNA. Nature *474*, 390-394.

Wang, K.C., and Chang, H.Y. (2011). Molecular mechanisms of long noncoding RNAs. Mol Cell *43*, 904-914.

Wang, K.C., Yang, Y.W., Liu, B., Sanyal, A., Corces-Zimmerman, R., Chen, Y., Lajoie, B.R., Protacio, A., Flynn, R.A., Gupta, R.A., *et al.* (2011b). A long noncoding RNA maintains active chromatin to coordinate homeotic gene expression. Nature *472*, 120-124.

Wang, P., Xue, Y., Han, Y., Lin, L., Wu, C., Xu, S., Jiang, Z., Xu, J., Liu, Q., and Cao, X. (2014). The STAT3-binding long noncoding RNA lnc-DC controls human dendritic cell differentiation. Science *344*, 310-313.

Wang, X., Arai, S., Song, X., Reichart, D., Du, K., Pascual, G., Tempst, P., Rosenfeld, M.G., Glass, C.K., and Kurokawa, R. (2008). Induced ncRNAs allosterically modify RNA-binding proteins in cis to inhibit transcription. Nature *454*, 126-130.

Wang, Y., He, L., Du, Y., Zhu, P., Huang, G., Luo, J., Yan, X., Ye, B., Li, C., Xia, P., *et al.* (2015). The long noncoding RNA lncTCF7 promotes self-renewal of human liver cancer stem cells through activation of Wnt signaling. Cell stem cell *16*, 413-425.

Wang, Y., Xu, Z., Jiang, J., Xu, C., Kang, J., Xiao, L., Wu, M., Xiong, J., Guo, X., and Liu, H. (2013). Endogenous miRNA sponge lincRNA-RoR regulates Oct4, Nanog, and Sox2 in human embryonic stem cell self-renewal. Developmental cell *25*, 69-80.

Ward, M., McEwan, C., Mills, J.D., and Janitz, M. (2015). Conservation and tissue-specific transcription patterns of long noncoding RNAs. Journal of human transcriptome *1*, 2-9.

Washietl, S., Kellis, M., and Garber, M. (2014). Evolutionary dynamics and tissue specificity of human long noncoding RNAs in six mammals. Genome Res *24*, 616-628.

Wassenegger, M., Heimes, S., Riedel, L., and Sanger, H.L. (1994). RNA-directed de novo methylation of genomic sequences in plants. Cell *76*, 567-576.

Waterston, R., and Sulston, J. (1995). The genome of Caenorhabditis elegans. Proc Natl Acad Sci U S A *92*, 10836-10840.

Waterston, R., Lindblad-Toh, K., Birney, E., Rogers, J., Abril, J.F., Agarwal, P., Agarwala, R., Ainscough, R., Alexandersson, M., An, P., *et al.* (2002). Initial sequencing and comparative analysis of the mouse genome. Nature *420*, 520-562.

Watters, K.M., Bryan, K., Foley, N.H., Meehan, M., and Stallings, R.L. (2013). Expressional alterations in functional ultra-conserved non-coding RNAs in response to all-trans retinoic acid-induced differentiation in neuroblastoma cells. BMC cancer *13*, 184.

Wei, J.T., Feng, Z., Partin, A.W., Brown, E., Thompson, I., Sokoll, L., Chan, D.W., Lotan, Y., Kibel, A.S., Busby, J.E., *et al.* (2014). Can urinary PCA3 supplement PSA in the early detection of prostate cancer? Journal of clinical oncology : official journal of the American Society of Clinical Oncology *32*, 4066-4072.

Wei, W., Pelechano, V., Jarvelin, A.I., and Steinmetz, L.M. (2011). Functional consequences of bidirectional promoters. Trends in genetics : TIG *27*, 267-276.

Weinberg, R.A., and Penman, S. (1968). Small molecular weight monodisperse nuclear RNA. Journal of molecular biology *38*, 289-304.

Werner, A. (2013). Biological functions of natural antisense transcripts. BMC biology *11*, 31.

Werner, M.S., and Ruthenburg, A.J. (2015). Nuclear Fractionation Reveals Thousands of Chromatin-Tethered Noncoding RNAs Adjacent to Active Genes. Cell reports *12*, 1089-1098.

Wery, M., Descrimes, M., Vogt, N., Dallongeville, A.S., Gautheret, D., and Morillon, A. (2016). Nonsense-Mediated Decay Restricts LncRNA Levels in Yeast Unless Blocked by Double-Stranded RNA Structure. Mol Cell *61*, 379-392.

Wightman, B., Ha, I., and Ruvkun, G. (1993). Posttranscriptional regulation of the heterochronic gene lin-14 by lin-4 mediates temporal pattern formation in C. elegans. Cell *75*, 855-862.

Willingham, A.T., and Gingeras, T.R. (2006). TUF love for "junk" DNA. Cell *125*, 1215-1220.

Wilusz, J.E., Freier, S.M., and Spector, D.L. (2008). 3' end processing of a long nuclear-retained noncoding RNA yields a tRNA-like cytoplasmic RNA. Cell *135*, 919-932.

Wong, L.H., Brettingham-Moore, K.H., Chan, L., Quach, J.M., Anderson, M.A., Northrop, E.L., Hannan, R., Saffery, R., Shaw, M.L., Williams, E., *et al.* (2007). Centromere RNA is a key component for the assembly of nucleoproteins at the nucleolus and centromere. Genome Res *17*, 1146-1160.

Wood, E.J., Chin-Inmanu, K., Jia, H., and Lipovich, L. (2013). Sense-antisense gene pairs: sequence, transcription, and structure are not conserved between human and mouse. Frontiers in genetics *4*, 183.

Wright, J.C., Mudge, J., Weisser, H., Barzine, M.P., Gonzalez, J.M., Brazma, A., Choudhary, J.S., and Harrow, J. (2016). Improving GENCODE reference gene annotation using a high-stringency proteogenomics workflow. Nat Commun *7*, 11778.

Wu, C.L., Wang, Y., Jin, B., Chen, H., Xie, B.S., and Mao, Z.B. (2015). Senescence-associated Long Non-coding RNA (SALNR) Delays Oncogene-induced Senescence through NF90 Regulation. The Journal of biological chemistry *290*, 30175-30192.

Wu, D.D., Irwin, D.M., and Zhang, Y.P. (2011). De novo origin of human protein-coding genes. PLoS Genet *7*, e1002379.

Xiang, J.F., Yin, Q.F., Chen, T., Zhang, Y., Zhang, X.O., Wu, Z., Zhang, S., Wang, H.B., Ge, J., Lu, X., *et al.* (2014). Human colorectal cancer-specific CCAT1-L lncRNA regulates long-range chromatin interactions at the MYC locus. Cell Res *24*, 513-531.

Xing, Z., Lin, A., Li, C., Liang, K., Wang, S., Liu, Y., Park, P.K., Qin, L., Wei, Y., Hawke, D.H., *et al.* (2014). lncRNA directs cooperative epigenetic regulation downstream of chemokine signals. Cell *159*, 1110-1125.

Xu, A.G., He, L., Li, Z., Xu, Y., Li, M., Fu, X., Yan, Z., Yuan, Y., Menzel, C., Li, N., *et al.* (2010). Intergenic and repeat transcription in human, chimpanzee and macaque brains measured by RNA-Seq. PLoS computational biology *6*, e1000843.

Xu, Z., Wei, W., Gagneur, J., Clauder-Munster, S., Smolik, M., Huber, W., and Steinmetz, L.M. (2011). Antisense expression increases gene expression variability and locus interdependency. Mol Syst Biol *7*, 468.

Xu, Z., Wei, W., Gagneur, J., Perocchi, F., Clauder-Munster, S., Camblong, J., Guffanti, E., Stutz, F., Huber, W., and Steinmetz, L.M. (2009). Bidirectional promoters generate pervasive transcription in yeast. Nature *457*, 1033-1037.

Yamazaki, T., and Hirose, T. (2015). The building process of the functional paraspeckle with long non-coding RNAs. Front Biosci (Elite Ed) *7*, 1-41.

Yan, X., Hu, Z., Feng, Y., Hu, X., Yuan, J., Zhao, S.D., Zhang, Y., Yang, L., Shan, W., He, Q., *et al.* (2015). Comprehensive Genomic Characterization of Long Non-coding RNAs across Human Cancers. Cancer cell *28*, 529-540.

Yang, F., Huo, X.S., Yuan, S.X., Zhang, L., Zhou, W.P., Wang, F., and Sun, S.H. (2013a). Repression of the long noncoding RNA-LET by histone deacetylase 3 contributes to hypoxia-mediated metastasis. Mol Cell *49*, 1083-1096.

Yang, L., Duff, M.O., Graveley, B.R., Carmichael, G.G., and Chen, L.L. (2011). Genomewide characterization of non-polyadenylated RNAs. Genome Biol *12*, R16.

Yang, L., Lin, C., Jin, C., Yang, J.C., Tanasa, B., Li, W., Merkurjev, D., Ohgi, K.A., Meng, D., Zhang, J., *et al.* (2013b). lncRNA-dependent mechanisms of androgen-receptor-regulated gene activation programs. Nature *500*, 598-602.

Yang, L., Li, L., Jiang, H., Shen, Y., and Plunkett, W. (2000). Expression of ERCC1 antisense RNA abrogates gemicitabine-mediated cytotoxic synergism with cisplatin in human colon tumor cells defective in mismatch repair but proficient in nucleotide excision repair. Clinical cancer research : an official journal of the American Association for Cancer Research *6*, 773-781.

Yao, H., Brick, K., Evrard, Y., Xiao, T., Camerini-Otero, R.D., and Felsenfeld, G. (2010). Mediation of CTCF transcriptional insulation by DEAD-box RNA-binding protein p68 and steroid receptor RNA activator SRA. Genes Dev *24*, 2543-2555.

Yildirim, E., Kirby, J.E., Brown, D.E., Mercier, F.E., Sadreyev, R.I., Scadden, D.T., and Lee, J.T. (2013). Xist RNA is a potent suppressor of hematologic cancer in mice. Cell *152*, 727-742.

Yin, Q.F., Yang, L., Zhang, Y., Xiang, J.F., Wu, Y.W., Carmichael, G.G., and Chen, L.L. (2012). Long noncoding RNAs with snoRNA ends. Mol Cell *48*, 219-230.

Yoon, J.H., Abdelmohsen, K., Kim, J., Yang, X., Martindale, J.L., Tominaga-Yamanaka, K., White, E.J., Orjalo, A.V., Rinn, J.L., Kreft, S.G., *et al.* (2013). Scaffold function of long non-coding RNA HOTAIR in protein ubiquitination. Nat Commun *4*, 2939.

Younger, S.T., Kenzelmann-Broz, D., Jung, H., Attardi, L.D., and Rinn, J.L. (2015). Integrative genomic analysis reveals widespread enhancer regulation by p53 in response to DNA damage. Nucleic Acids Res *43*, 4447-4462.

Yu, W., Gius, D., Onyango, P., Muldoon-Jacobs, K., Karp, J., Feinberg, A.P., and Cui, H. (2008). Epigenetic silencing of tumour suppressor gene p15 by its antisense RNA. Nature *451*, 202-206.

Yuan, C., Wang, J., Harrison, A.P., Meng, X., Chen, D., and Chen, M. (2015). Genome-wide view of natural antisense transcripts in Arabidopsis thaliana. DNA research : an international journal for rapid publication of reports on genes and genomes *22*, 233-243.

Yuan, J.H., Yang, F., Wang, F., Ma, J.Z., Guo, Y.J., Tao, Q.F., Liu, F., Pan, W., Wang, T.T., Zhou, C.C., *et al.* (2014). A long noncoding RNA activated by TGF-beta promotes the invasion-metastasis cascade in hepatocellular carcinoma. Cancer cell *25*, 666-681.

Yuan, S.X., Yang, F., Yang, Y., Tao, Q.F., Zhang, J., Huang, G., Wang, R.Y., Yang, S., Huo, X.S., Zhang, L., *et al.* (2012). Long noncoding RNA associated with microvascular invasion in hepatocellular carcinoma promotes angiogenesis and serves as a predictor for hepatocellular carcinoma patients' poor recurrence-free survival after hepatectomy. Hepatology *56*, 2231-2241.

Yue, F., Cheng, Y., Breschi, A., Vierstra, J., Wu, W., Ryba, T., Sandstrom, R., Ma, Z., Davis, C., Pope, B.D., *et al.* (2014). A comparative encyclopedia of DNA elements in the mouse genome. Nature *515*, 355-364.

Zhang, A., Zhao, J.C., Kim, J., Fong, K.W., Yang, Y.A., Chakravarti, D., Mo, Y.Y., and Yu, J. (2015). LncRNA HOTAIR Enhances the Androgen-Receptor-Mediated Transcriptional Program and Drives Castration-Resistant Prostate Cancer. Cell reports *13*, 209-221.

Zhang, B., Arun, G., Mao, Y.S., Lazar, Z., Hung, G., Bhattacharjee, G., Xiao, X., Booth, C.J., Wu, J., Zhang, C., *et al.* (2012). The lncRNA Malat1 is dispensable for mouse development but its transcription plays a cis-regulatory role in the adult. Cell reports *2*, 111-123.

Zhang, B., Gunawardane, L., Niazi, F., Jahanbani, F., Chen, X., and Valadkhan, S. (2014). A novel RNA motif mediates the strict nuclear localization of a long noncoding RNA. Mol Cell Biol *34*, 2318-2329.

Zhang, X., Lian, Z., Padden, C., Gerstein, M.B., Rozowsky, J., Snyder, M., Gingeras, T.R., Kapranov, P., Weissman, S.M., and Newburger, P.E. (2009). A myelopoiesis-associated regulatory intergenic noncoding RNA transcript within the human HOXA cluster. Blood *113*, 2526-2534.

Zhang, Y., He, Q., Hu, Z., Feng, Y., Fan, L., Tang, Z., Yuan, J., Shan, W., Li, C., Hu, X., *et al.* (2016). Long noncoding RNA LINP1 regulates repair of DNA double-strand breaks in triple-negative breast cancer. Nat Struct Mol Biol *23*, 522-530.

Zhang, Y., Zhang, X.O., Chen, T., Xiang, J.F., Yin, Q.F., Xing, Y.H., Zhu, S., Yang, L., and Chen, L.L. (2013). Circular intronic long noncoding RNAs. Mol Cell *51*, 792-806.

Zhao, J., Sun, B.K., Erwin, J.A., Song, J.J., and Lee, J.T. (2008). Polycomb proteins targeted by a short repeat RNA to the mouse X chromosome. Science *322*, 750-756.

Zheng, D., Frankish, A., Baertsch, R., Kapranov, P., Reymond, A., Choo, S.W., Lu, Y., Denoeud, F., Antonarakis, S.E., Snyder, M., *et al.* (2007). Pseudogenes in the ENCODE regions: consensus annotation, analysis of transcription, and evolution. Genome Res *17*, 839-851.

Zheng, G.X., Do, B.T., Webster, D.E., Khavari, P.A., and Chang, H.Y. (2014). Dicer-microRNA-Myc circuit promotes transcription of hundreds of long noncoding RNAs. Nat Struct Mol Biol *21*, 585-590.

Zheng, R., Shen, Z., Tripathi, V., Xuan, Z., Freier, S.M., Bennett, C.F., Prasanth, S.G., and Prasanth, K.V. (2010). Polypurine-repeat-containing RNAs: a novel class of long non-coding RNA in mammalian cells. Journal of cell science *123*, 3734-3744.

Zheng, S., Vuong, B.Q., Vaidyanathan, B., Lin, J.Y., Huang, F.T., and Chaudhuri, J. (2015). Non-coding RNA Generated following Lariat Debranching Mediates Targeting of AID to DNA. Cell *161*, 762-773.

Zhou, Y., Zhong, Y., Wang, Y., Zhang, X., Batista, D.L., Gejman, R., Ansell, P.J., Zhao, J., Weng, C., and Klibanski, A. (2007). Activation of p53 by MEG3 non-coding RNA. The Journal of biological chemistry *282*, 24731-24742.

Ziegler, A., Luedke, G.H., Fabbro, D., Altmann, K.H., Stahel, R.A., and Zangemeister-Wittke, U. (1997). Induction of apoptosis in small-cell lung cancer cells by an antisense oligodeoxynucleotide targeting the Bcl-2 coding sequence. Journal of the National Cancer Institute *89*, 1027-1036.

Zieve, G., and Penman, S. (1976). Small RNA species of the HeLa cell: metabolism and subcellular localization. Cell *8*, 19-31.

Zucchelli, S., Fasolo, F., Russo, R., Cimatti, L., Patrucco, L., Takahashi, H., Jones, M.H., Santoro, C., Sblattero, D., Cotella, D., *et al.* (2015). SINEUPs are modular antisense long non-coding RNAs that increase synthesis of target proteins in cells. Frontiers in cellular neuroscience *9*, 174.

Glossary

Amino acid: a carboxylic acid that also contains a functional amine group. Such organic compounds therefore have a carboxyl group –COOH and an amine group –NH2 or –NH–. 149 different amino acids are present in proteins.

Actinomycin D: peptide antibiotic isolated from Streptomyces bacteria. Besides its therapeutic properties, actinomycin D also blocks the transcription of DNA into messenger RNA with RNA polymerase by binding beside the promoter.

Chromatin remodeler: a complex that enables the modification of chromatin's structure using energy freed through ATP hydrolase. Several proteins are involved in this remodeling process (SW1/SNF, ISW1, INO, CHD families). Control of the chromatin structure also allows regulation of gene expression.

Cryptic peptides: same notion as for RNA where the synthesized peptide is not measurable as its half-life is so short.

Cryptic RNA: a "hidden" RNA, often due to its very short half-life.

Differential hybridization: term defining the difference in hybridization capacity of the same oligonucleotide probe on DNA or RNA from different conditions. This term was largely used for DNA or RNA chips or the quantification of hybridization assays (covering the whole genome) obtained through fluorescent markers.

Endonuclease: unlike exonuclease, this nuclease cuts within the RNA or DNA molecule (following a template or a defined distance).

Enhancers: fundamental *cis*-regulator elements for the control of gene expression, providing a platform for the recruitment of transcription factors and the establishment of specific chromatin organization. Many, if not all, functional enhancers are transcribed in a permissive manner, in both directions (sense and antisense) and are called eRNA. Transcribed enhancer regions have specific signatures at the histone level that distinguish them from other transcription units.

Epigenetics: term that defines the transmissible and reversible modifications, in the absence of the original signal, of gene expression (and function) that are not followed by nucleotide sequence changes. DNA modification and changes in chromatin (structure, position, modifications, and compaction) in particular are classed as epigenetic modifications whose code, not yet clear, has been suggested.

Exonuclease: a nuclease enzyme that cuts nucleic acids (DNA or RNA). The *exo* prefix shows that the cleavage occurs at the extremities, one nucleotide at a time, in the 5' to 3' sense direction or the other way.

GENCODE portal: aims to identify all the characteristics of genes in the human genome by using a combination of computer analyses, manual labeling and experimental validation.

Gene: unit that defines a region containing transcription start and end signals. Originally, a gene was thought to contain information from the genetic code for the expression of a protein. The term is now also applied to regions that do not code for proteins.

Genetic code: represents all the rules that enable the translation of information held within the genome for protein synthesis. It provides the link between the genotype and phenotype of an organism. This code relies on nucleotide triplets (codons), messenger RNA and the amino acids incorporated in the synthesized proteins during translation of messenger RNA by ribosomes.

Heterologous expression: describes protein or RNA expressed from a promoter other than its original promoter and/or from another chromosomal/plasmid region and/or in another organism.

Histones: proteins located in the eukaryote cell nucleus and in archaebacteria. They are the main protein constituents of chromosomes and are strongly associated with DNA whose compaction they enable by forming the nucleosome.

HOX genes: code for a family of conserved transcription factors that regulate the spatial organization of the embryonic body and that contribute to the cell specification of several differentiation processes. In mammals, there are 39 HOX genes that are classified into four groups (HOXA, HOXB, HOXC and HOXD), with a precise spatio-temporal coordination of their respective expressions.

Insulator: DNA sequences sometimes found between two genes or groups of genes, thus protecting them from the effects produced by other regulator sequences. These elements act as genetic barriers preventing interaction between enhancer/amplifier elements and the promoter. They can bind protein factors such as CTCF to form a physical barrier.

Interleukin: a type of cytokine of the immune system contributing the organism's response to microbial infection (increases lymphocyte proliferation) and to the auxiliary T lymphocyte response.

Locus 8q24: discovered in 2006, this was one of the first loci associated with a genetic risk to be validated in a reproducible way. Since then, it has also been confirmed for other cancers such as colon, breast and bladder cancer. The exact mapping of the locus revealed alleles covering distances of nearly 490 kilobases. Curiously, risk variants are found in the non-coding region and the mechanism by which they contribute to cancers remains unknown. These alleles are close to the MYC oncogene and results suggest one or more interconnected roles between these non-coding RNA and MYC as the risk factor.

Mediator: a multiprotein complex that acts as a coactivator of transcription in eukaryotes. It was discovered in the laboratory of Roger D. Kornberg, the 2006 Nobel Prize in Chemistry laureate. The mediator interacts with transcription factors and RNA polymerase II. Its main function is to transmit transcription factor signals towards the polymerase.

Nucleosome: composed of a dimer formed from the 4 histones H3, H2A, H2B and H4, around which 147 DNA base pairs are wrapped.

Permissive transcription: represents an RNA transcription event in which the DNA structure is weak or does not lead to RNA that is competent for translation and therefore rapidly degrades.

Permissive translation: same notion as for translation but where the RNA reading by the ribosome does not lead to a stable or functional protein or a protein of insufficient size.

Pioneer translation: represents the first "round" of translation by one or several ribosomes of a messenger RNA. This first cycle is considered as a stage necessary for the establishment of robust translation of an mRNA molecule.

Polycomb: polycomb proteins were discovered for the first time in *Drosophila melanogaster* and can remodel chromatin to switch off gene expression. For example, HOX genes are mainly regulated by polycomb factors during development. It is now clear that these factors are themselves controlled, targeted or activated by non-coding RNA in a genome region-specific manner.

Pre-genomic era: a period in the history of molecular biology during which no genomes had yet been characterized and/or techniques for measuring the whole genome had not yet been fully developed.

Promoter: sequence constituting the basic unit for the recruitment of transcription machinery to which transcription factors and the transcription preinitiation complex (PIC) bind. Many promoters are activated through enhancers situated at mid to long distances upstream or downstream of them.

Riboswitches: RNA structures found on the 5' end, untranslated, of some messenger RNA. They are composed of two parts: the aptamer and the expression platform. The aptamer can bind directly to a ligand which initiates modification of the structure of the expression platform and affects the expression of the mRNA-carried gene by blocking or activating the transcription or translation of the corresponding protein.

Ribozyme: RNA that have catalytic properties for a specific chemical reaction. They are considered as contemporaries of **Terminator** enzymes; sequences that contain the information necessary for transcription termination. In mammals, they contain RNA cleavage and polyadenylation signals during synthesis, initiating the end and release of RNA polymerase.

RNA world: the hypothesis of an RNA world puts forward the idea that ribonucleic acid was the precursor of all biological macromolecules, especially DNA and proteins. This hypothesis provides and explanation for the development and appearance of different biological functions in studies on the origins of life. Walter Gilbert first used the term RNA World in 1986.

Shine-Dalgarno: a ribosomal binding site (consensus sequence AGGAGG) on messenger RNA of prokaryotes and archae, located at around 8 bases upstream of the START AUG codon. It therefore allows the ribosome to orient itself for a correct initiation. It bears the name of its discoverers John Shine and Lynn Dalgarno.

Splicesosome: called a splicing particle, this is a dynamic complex of ribonucleoprotein particles (composed of RNA and over 200 proteins) found in the cell nucleus.

Syntenic loci: gene regions whose chromosomal organization is similar across related species without there being any genetic link or sequence conservation. These are used today as tools to determine whether lncARN whose sequence is not conserved has a conserved function based on its genomic position in different species.

Transgene: the isolated sequence of a gene, transferred from one organism to another through the implementation of transgenesis.

List of Acronyms

Air/Airn	*Antisense Igf2r RNA*
Alu	Sub-family of repeat SINE elements
ANRIL	*Antisense Non-coding RNA in the INK4 Locus*
Antago-NAT	Small RNA molecule that targets the antisense strand of a target gene: *ANTAGOnist Natural Antisense Transcript*
arcRNA	Family of non-coding architectural RNA
ARG1-AS	Antisense RNA of the *ARG1* gene (yeast) coding for *arginine methyl transferase 1*
ASncmtRNA-2	Antisense non-coding mitochondrial RNA 2
asRNA	Antisense RNA
AS-UChl1	Antisense RNA for the *Uchl1* gene
Ataxin-1	Protein containing a DNA binding site and with a role in the spinocerebellar ataxia type 1 neurodegenerative disease
ATPase	ATP hydrolase
BACE1	*Beta-secretase 1* gene
BACE1-AS	Long non-coding antisense RNA of BACE1
BCAR4	Long non-coding RNA *Breast Cancer Anti-Estrogen Resistance 4* promotes proliferation
BCL2	Gene coding for *B-cell lymphoma 2* protein to regulate cell death
BCL2/IGH	Chimeric RNA of BCL2 and IGH
BDNF	*Brain-Derived Neurotrophic Factor* gene

BMI-1	B lymphoma Mo-MLV insertion region 1 homolog, member of the polycomb complex
BRG1	Brahma related gene 1, codes for the catalytic subunit of the Swi-Snf complex
Bvht	Long non-coding RNA *braveheart*
CAR	Family of non-coding RNA *Chromatin Associated lncRNA*
CARLo-5	Cf CCAT1
CBP/p300	Complex containing 2 transcriptional coactivators *CREB Binding Protein* and *P300* able to bind and modify histones (histone acetylation domain)
CCAT1/CARLo-5	Long non-coding RNA *Colon Cancer Associated Transcript 1*
CCAT2	Long non-coding RNA *Colon Cancer Associated Transcript 2*
CCND1	*Cyclin D1* gene
CDKN1A/p21	*Cyclin Dependent Kinase Inhibitor 1A*
CDKN2A/p16	*Cyclin Dependent Kinase Inhibitor 2A,* tumor suppressor
CDKN2B/p15	*Cyclin Dependent Kinase Inhibitor 2B,* tumor suppressor
cDNA	complementary DNA
CDR1	*Cerebellar degeneration-related protein 1* gene
CDR1as	Non-coding antisense gene of CDR1
CENP-A/C	*Centromeric protein A* and *C* partially form the kinetochore (protein assembly of centromeric regions in chromosomes)
cheRNA	Chromatin enriched non-coding RNA
CHiP	*CHromatin ImmunoPrecipitation*
circRNA/cirRNA	Circular RNA (non-coding RNA)
ciRS-7/CDR1AS	Transcribed (circular) antisense of the *CDR1* gene
COREST	*COfactor of Repressor Element-1 Silencing Transcription factor*
COX2	*Cyclooxygenase-2* gene
CryoEM	*Cryo Electro-Microscopy*
CTCF	Gene coding for *CCCTC-binding Factor*

CUT	Class of non-coding RNA in *cerevisiae Cryptic Unstable Transcript*
Cyrano/OIP5-AS	Long non-coding RNA of the zebrafish involved in brain development
DANCR	Differentiation Antagonizing Non-protein Coding RNA
DBE-T	Long non-coding RNA *D4Z4 binding element transcript*
DICER	Protein that cuts the double strands of RNA into small regulator RNA molecules
DISC	Locus containing genes involved in schizophrenia and containing long non-coding antisense RNA
Dlx1os	Long antisense RNA *Distal less homeobox 1 opposite strand 1*
Dlx6	Gene coding for the homeobox 6 protein, important for brain development
Dlx6OS1	Long non-coding RNA *Distal less homeobox 6 opposite strand 1*
DNA	Deoxyribonucleic Acid
DNAse	*DNA nuclease*
DROSHA	Member of ribonuclease 3 family that cleaves the 3' and 5' extremities of stem and loop RNA to form pre-miARN
DWORF	*Dwarf Open Reading Frame*, long non-coding RNA that contains the sequence of a small peptide involved in muscular contraction in mice
Dzip3	*DaZ Interacting Protein 3*, type 3 E3 ubiquitine ligase that interacts with the HOTAIR lncRNA
E	*Enhancer region*
E2F1	Human transcription factor that binds to retinoblastoma promoters
E3	Type of ubiquitine ligase that catalyzes the linkage of charged ubiquitine on the donor of E2 protein towards the protein substrate targeted for degradation.
EHMT2	Gene coding for *Euchromatic Histone Lysine Methyltransferase 2*, responsible for the monomethylation and dimethylation of the H3 histone to lysine 9 (H3K9me1 and me2)
ENCODE	*ENCyclOpedia of DNA Elements*

endo-si-lncRNA	Family of long non-coding RNA, precursor of siRNA
EPHB2	Gene coding for *EPHrin receptor B2*
eRNA ou RNAe	Family of non-coding enhancer RNA, enhancer RNA
ES cells	Embryonic Stem cells (ESc)
Evf2	Long non-coding RNA in mice equivalent to Dlx6as (human)
EZH2	Gene coding for *Enhancer of Zeste Homolog 2*, member of the polycomb 2 complex and catalyst for the methylation of the H3 histone to lysine 27 (H3K27me)
FAL1	Long non-coding RNA *Focally Amplified Long Noncoding on Chromosome 1*
FANTOM	Consortium *Functional ANnoTatiOn of the Mammalian genome*
Fendrr	Long non-coding RNA, adjacent regulator of Foxf1
FOXC1	*Forkhead Box C1* gene
FOXC1e	*FOXC1 enhancer*
Foxf1	*Forkhead box protein F1* gene
Frigidair	Long non-coding antisense RNA of HOXC, regulates HOTAIR
Ftx	Long non-coding RNA that regulates XIST on the X chromosome
Gad1	Gene coding for *glutamic acid decarboxylase 1,* expressed in the brain
GADD45a	Gene coding for *Growth Arrest and DNA Damage inducible Alpha*
GAS5	Long non-coding RNA *Growth-Arrest-Specific transcript 5*
GDNF	Gene coding for Glial- Derived Neurotrophic Factor
GRSF1	Gene coding for *G-Rich RNA Sequence Binding Factor 1*
H19	Long non-coding RNA expressed through the maternal chromosome 11 in humans
H3-K27	Histone H3 lysine 27
H3-K4	Histone H3 lysine 4
H4K20	Histone H4 lysine 20

HAR	*Human Accelerated Region*
HARIA	Long non-coding RNA, specific to humans, *Highly Accelerated Region IA*
HeLa	Human immortal cells *Henrietta Lack*
HGP	*Human Genome Project*
HINCUT	Long non-coding *Hypoxia-Induced NonCoding Ultraconserved Transcript*
HJURP	Gene coding for *Holliday JUnction Recognition Protein,* a chaperone involved in depositing the centromeric histone CENPA
HLXB9-LncRNA	Long non-coding RNA that overlaps with the coding gene for the embryonic transcription factor HLXB9
hnRNP-K	Gene coding for *heterogeneous nuclear ribonucleoprotein K,* for nucleic acid binding
HOTAIR	Long non-coding antisense RNA of HOXC, *HOX transcript antisense RNA*
HOTAIRM1	Long non-coding RNA *HOXA Transcript AntIsense RNA, Myeloid-Specific 1*
HOTTIP	Long non-coding antisense RNA of HOXA, *HOXA Distal Transcript Antisense RNA*
HuR	*HUman antigen R* gene
ICR	*Imprinting Controlling Region*
IGF2R	*Insulin Like Growth Factor 2 Receptor* gene
IGH	*ImmunoGlobulin Heavy locus* gene whose rearrangements with other genes is apparent in several cancers
IkB	Gene coding for a kinase with an important role in inflammation
IL1β-eRNA	Long non-coding enhancer RNA of the gene IL1-B coding for *interleukin 1-Beta*
INK4A/ARF	Locus containing several genes that code for tumor repressors and non-coding genes, found on chromosome 9p
iPSC	*Inducible Pluripotent Stem Cell*
Kb	Kilo base (1000 bases)
Kcnq1ot1	Long non-coding RNA overlapping Kcnql
Kcnql	*K+ channel* gene

Khps1	Long anti sense RNA from Proto-oncogene SPHK1, encoding *sphingosine kinase 1* involved in EMT
KSRP	K-type splicing regulator protein
LED	Long non-coding RNA: *LncRNA activator of Enhancer Domains*
LET	Long non-coding RNA: *Low Expression in Tumor* invasioninvasion inhibitor
Let-7	miARN *lethal-7,* involved in regulating development in *C. elegans*
Lethe	Long non-coding RNA that regulates the inflammatory response
lin-14	Gene coding for a transcription factor in *C. elegans* and involved in embryonic development
Lin-4	Micro-RNA (miRNA) regulating the expression of the Lin-14 gene
Linc-MD1	Long non-coding RNA *Long intergenic non-protein coding RNA, Muscle Differentiation 1* involved in muscular differentiation
lincRNA	Family of intergenic long non-coding RNA, *Long Intergenic Non-Coding RNA (lincRNA)*
LincRNA-EPS	Long non-coding murine RNA *LincRNA Erythroid ProSurvival*
LincROR	Long non-coding RNA *Long intergenic Noncoding RNA, Regulator Of Reprogramming,* involved in pluripotency and induced during hypoxic stress (oxygen deficiency)
LINP1	Long non-coding RNA involved in DNA repair: *LncRNA In Non-Homologous End Joining Pathway 1*
Linx	Long non-coding RNA close to XIST
Lnc-DC	Long non-coding RNA expressed in dendritic cells, *Long non -coding-Dendric cells*
lncRNA	Family of long non-coding RNA, *Long Non-Coding RNA (lncRNA)*
lncRNA-AA174084	Long non-coding RNA expressed in gastric cancer
LncRNA-ATB	Long non-coding RNA *LncRNA-Activated by TGF-β*
lncRNA-CCND1	Long non-coding RNA upstream of the promoter *CCND1*

LncTCF7	Long non-coding RNA that promotes the proliferation of cancers
LnX3	*Ligand of Numb-protein X 3* gene, contains XIST sequence homology and is its likely precursor
Lrp	*Leucine Regulator Protein*
LSD1	*LySine Demethylase 1* gene, member of the CoREST/REST complex
LTR	*Long Terminal Repeat*
LUAT	Family of long non-coding RNA upstream of *Long Upstream Antisense Transcript* genes
LUNAR1	Long non-coding RNA activated by NOTCH, *LeUkemia-induced Noncoding Activator RNA*
MALAT1	Long non-coding RNA *Metastasis Associated Lung Adenocarcinoma Transcript 1*
MAML1	*Mastermind Like Transcriptional Coactivator 1* gene, involved in the Notch channel (intercellular signaling channel)
MEF2c	*Myocyte Enhancer Factor 2C* gene, involved in myogenesis.
MEG3	Long non-coding RNA *Maternally Expressed Gene 3*
Mex3B	Gene coding for a type 3ubiquitin ligase, facilitating the degradation of snurportin
micF	antisense RNA regulator of OmpF
mi-lncRNA	Family of long non-coding RNA, precursor of miRNA
MIR31HG	Long non-coding RNA involved in the regulation of P16 and containing miRNA31
miR-675-3P	miRNA from the H19 non-coding RNA
miR-7	miRNA bonded to a circular non-coding RNA
Mira/Mistral	Long non-coding RNA expressed on the HOX locus in mice
miRNA	Family of non-coding RNA, micro RNA
MLL1	*Mixed-Lineage Leukemia 1* gene, catalyzes the methylation of the H3 histone in lysine 4
MUT	Family of non-coding RNA in *S. cerevisiae, Meiotic Unstable Transcript*
MYC	Proto-oncogenic transcription factor
NANOG	Gene coding for the Homeobox transcription factor, involved in the proliferation of stem cells

NAT	Family of non-coding RNA *Natural Antisense Transcript*
NBAT-1	Long non-coding RNA *NeuroBlastoma Associated Transcript 1*
ncmtRNA	Non-coding mitochondrial RNA
ncRNA	Non-coding RNA
ncRNAa	Non-coding RNA activating
NEAT1	Long non-coding RNA *Nuclear Paraspeckle Assembly Transcript 1*
NER	DNA repair process: Nucleotide Excision Repair
NeST	Long non-coding RNA activating *Nettoie Salmonella pas Theiler's*
NF90	*Nuclear Factor 90* gene
NFKB	Alternative name of the pseudogene of the ribosomal protein S15a (Rsp15a)
NF-kB	Gene coding for the *Nuclear Factor-kappa Beta* transcription factor
NIH	*National Institute of Health*
NKILA	Long non-coding RNA *NF-KappaB Interacting LncRNA*
NMD	RNA degradation channel: *Non-sense Mediated Decay*
NORAD	Long non-coding RNA that binds to Pumillo, *NOn-coding RNA Activated by DNA Damage*
Nrd1	Gene coding for the *Nuclear pre-mRNA Down-Regulation 1* protein
NRIP1	*Nuclear Receptor Interacting Protein 1* gene
NRIP1e	*NRIP1 enhancer*
NRSF	Cf REST
nt	Nucleotide
NUT	Family of non-coding RNA in *cerevisiae Nrd1-Unstable Transcript*
OCT4	Gene coding *OCtamer-binding Transcription factor 4*, involved in pluripotency
ODN	*OligoDeoxyNucleotides* used for RNA silencing by complementarity
OIP5	*Opa Interacting Protein* gene
OIP5-AS1	Long non-coding antisense RNA of OIP5 (Cyrano in Zebrafish)

OmpF	*Outer Membrane Porin gene F* gene
p14ARF	Isoform of CDKN2A
P15-As/ANRIL	Long antisense RNA of p15INK4b, *Antisense Noncoding RNA in the INK4 Locus*
p15INK4b	Cf CDKN2B
p16INK4a	Cf CDKN2A
P250	Gene coding for the general transcription factor *TBP-associated factor 250 kDa* or TAF1
P53	Transcription factor involved in tumor development, *Tumor protein p53*
PALR	*Promoter Associated Long non-coding RNA*
pancRNA	Family of non-coding RNA associated with gene promoters
PANDA	Long non-coding RNA close to CDKN1A, regulates apoptosis
PAPAs	Non-coding RNA from the promoter and antisense of rRNA
paRNA	Promoter-associated RNA
PCA3	Long non-coding RNA *Prostate CAncer gene 3,* diagnostic marker of prostate cancer
PCA3	Long non-coding RNA *Prostate Cancer 3* expressed in cancerous prostate tissues
PCAT	Family of non-coding RNA, *Prostate Cancer Associated Transcript*
PCG	*Promoter of Coding Gene*
PCGEM1	Long non-coding RNA expressed in prostate cancer
PHO84-CUT	CUT-type long non-coding RNA, antisense of the PHO84 gene (involved in phosphate metabolism) in *S. cerevisiae*
PIC	*Promoter Initiation Complex*
pi-lncARN	Family of long non-coding RNA, precursor of piRNA
piRNA	*piwi-associated RNA*
PIWI	Gene coding for the PIWI protein in the Argonaut family that interacts with piRNA
PNUTS	*Protein phosphatase 1 (PP1) NUclear Targeting Subunit* gene involved in apoptosis
Poldi	Gene coding for *de novo* in mice, from a non-coding gene

Poly-A	Poly(A)+
POT1	*Protection Of Telomeres 1* gene
PRC1 and **PRC2**	*Polycomb repressive Complex 1* and *2*
PROMPT	Family of long non-coding RNA *PROMoter uPstream Transcript*
PSA	*Prostate Specific Antigen* gene
PTENP1	Long non-coding RNA (circular) of *PTEN tumor suppressor Pseudogene 1*
PTGS	Post Transcriptional Gene Silencing
PUMILLO	Protein that binds to RNA and regulates genomic stability
PVT1	Long non-coding RNA *Plasmacytoma Variant Translocation 1*
PYCARD	Gene coding for an apoptosis factor
PYD	Protein region containing a *Pyrin Domain (PYD)*, important for caspase activity during apoptosis
RBS	Ribosome Binding Site
REST	*Repressor Element-1 Silencing Transcription factor*, recognizes a 24-base pair sequence in DNA and recruits multiple epigenetic factors
R-fold	Program used to predict RNA structure
RIKEN	Japanese research institute
RISC	*RNA-Induced Silencing Complex*
RME2	Long non-coding RNA (*cerevisiae*) *Regulator of Meiosis 2*
RMRP	Gene coding for *RNA Component of Mitochondrial RNA Processing* endoribonuclease
RNA	Ribonucleic Acid
RNA-GRC	GAA repeat-containing RNA
RNAi	RNA interference
RNAPII, RNAPIII RNAPIV and **V**	RNA polymerase II, III, IV and V
RnaseP	Type P Ribonuclease
RNA-seq	RNA sequencing
RnaseZ	Type Z Ribonuclease
RNP	*RiboNucleo Particle*

Rox2	*RNA On the X2*, non-coding RNA involved in the X chromosome activation in drosophila
RPA	Gene coding for *Replication Protein A* involved in telomere protection
rRNA	ribosomal RNA
Rrp6	Gene coding for *Ribosomal RNA Processing 6,* RNA exonuclease 3'-5' and part of the exosome.
RTL	Long antisense RNA of the LTR repeat element (Long Terminal Repeat) of the transposable Ty1 element
SAFA	*Scaffold-Attachment-Factor* associated with PANDA for the regulating the senescence gene
SAGE	Serial Analysis of Gene Expression
SAL	Family of non-coding RNA activated in senescence
SALNR	Long non-coding RNA *Senescence Activated Long Noncoding RNA*
SAMMSON	Long non-coding RNA expressed in human melanomas
SchLAP1	Long non-coding RNA expressed in prostate cancer: *SWI/SNF Complex Antagonist associated with Prostate cancer 1*
SD	Shine Dalgarno
SHAPE	Selective 2'-Hydroxyl acylation Analyzed by Primer Extension: RNA structure analysis method
Si-lncRNA	Stress-induced lncRNA
SINEB2	Transposable element in mice (*Short INterspaced Element B2*)
SINE-UP	Long antisense RNA in the 5' region of genes containing a SINE repeat element in mice
sisR-1	Intronic long non-coding regulator sis-type RNA in drosophila
sisRNA	Family of long non-coding RNA *Stable Intronic Sequence RNA*
SMAD	Family of genes coding for transcription factors (*Similarity to the Drosophila gene Mothers Against Decapentaplegic*)
SNIP1	*Smad Nuclear Interacting Protein 1* gene
SNP	Single Nucleotide Polymorphism

snRNA	Small nucleic RNA
Snurportin-1	*Nucleoplasmic shuttling protein* gene
SOX2-OT	Long non-coding RNA that overlaps with the SOX2 gene: *SOX2 overlapping transcript*
SPHK1	*Sphingosine kinase 1* gene (cf lncARN KHPS1)
SRA	Long non-coding RNA *Steroid Receptor RNA Activator*
sRNA	Family of small regulator RNA
STAT3	Gene coding for the Signal Transducer and Activator of Transcription 3, involved in the activation of many stress genes.
STAU1	*STAUfen double-stranded RNA-binding protein 1* gene, involved in the transport and degradation of RNA
SUT	Family of long non-coding RNA in *cerevisiae Stable Uncharacterized Transcript*
Suv4-20h2	Gene coding for the methyl transferase of histone H4 in lysine 20
SUZ12	*SUppressor of Zeste 12* member of the PRC2 complex
SWI/SNF-like	Chromatin remodeling complex: *SWItch/Sucrose Non-Fermentable*
TAFI, TAFII	Transcription factor 1 and 2
TCF21	Tumor suppressor transcription factor, TransCription Factor 21
TCF7	Tumor suppressor transcription factor, TransCription Factor 7
TE	Transposable Element
TERC	Long non-coding RNA that catalyzes the elongation of telomeres *TElomerase RNA Component*
TERRA	Long non-coding TElomeric Repeat RNA containing
TERT	*TElomerase Reverse Transcriptase* gene involved in the elongation of telomeres
TGFB	*Transforming Growth Factor B*
TGS	*Transcriptional Gene Silencing*
TINCR	Long non-coding RNA *Tissue Differentiation-Inducing Non-Protein Coding RNA*
tRNA	Transfer RNA

Tsix	Long non-coding antisense RNA of XIST
TSS	Transcription Start Site
TTS	Transcription Termination Site
T-UCR	Family of non-coding RNA *Transcript from Ultraconserved Region*
TUG1	Long non-coding RNA *Taurine Upregulated Gene 1*
uaRNA	upstream antisense RNA
Uc. 283+A	Long non-coding UCR-T type RNA involved in the regulation and maturation of miARN in humans
Uchl1	*Ubiquitin C-Terminal Hydrolase L1* gene
UCR	Ultra-Conserved Regions
Unc-22	*UNCoordinated locomotion protein 22*, or twitchin in *c. elegans*
UTR	UnTranslated Region
Vlinc-/macro-lncRNA	Very long intergenic non- coding RNA
XACT	Human *X-ACtivating Transcript*
XCI	X Chromosome Inactivation
XIC	X Inactivation Center
Xist	Long non-coding RNA *X-Inactive Specific Transcript*
Xrn1	Exoribonuclease 5'-3'
XUT	*Xrn1-sensitive Uncharacterized Transcript*
YY1	*Yin Yang 1* transcription factor involved in the activation and repression of genes
ZEB2	Gene coding for *Zinc Finger E-Box Binding Homeobox 2 protein*
ZEB2NAT	Long antisense RNA of the ZEB2 gene, *ZEB2 Natural Antisense Transcript*

Index

FSC
www.fsc.org
MIX
Papier | Fördert
gute Waldnutzung
FSC® C083411

Kolibri 360 GmbH
Ferdinand-Jühlke-Straße 7
99095 Erfurt, Deutschland
produktsicherheit@kolibri360.de